Proceedings in Life Sciences

Insect Life History Patterns: Habitat and Geographic Variation

Edited by
Robert F. Denno and Hugh Dingle

With 62 Figures

Springer-Verlag
New York Heidelberg Berlin

Robert F. Denno
Department of Entomology
University of Maryland
College Park, Maryland 20742, U.S.A.

Hugh Dingle
Department of Zoology
University of Iowa
Iowa City, Iowa 52242, U.S.A.

Production: Kate Ormston

The figure on the front cover is a stylized drawing (after a photograph by Tom Woods) of a thornbug, *Umbonia crassicornis,* which feeds on leguminous trees.

Library of Congress Cataloging in Publication Data
Main entry under title:
Insect life history patterns.
(Proceedings in life sciences)
Bibliography: p.
Includes index.
1. Insect-plant relationships. 2. Insects–Variation. I. Denno, Robert F. II. Dingle, Hugh, 1946- . III. Series.
QL463.17 595.7'05 81-1576
AACR2

Printed in the United States of America

9 8 7 6 5 4 3 2 1

ISBN 0-387-90591-X Springer-Verlag New York Heidelberg Berlin
ISBN 3-540-90591-X Springer-Verlag Berlin Heidelberg New York

Preface

This volume results from a symposium entitled "Species and Life History Patterns: Geographic and Habitat Variation", held during the National Meeting of the Entomological Society of America in Denver, Colorado, USA in November, 1979. The stimulus to assemble papers on this theme emerged from continuing discussions with colleagues concerning controversies in ecology and evolutionary biology, namely those associated with plant-herbivore interactions, life history theory, and the equilibrium status of communities. The study organisms used in this series of reports are all either herbivorous insects or those intimately associated with plants. In this volume we stress the variation found in life history traits and address some of the problems inherent in current life history theory.

We include as life history traits not only traditional variables such as fecundity, size of young, and age to first and peak reproduction, but also diapause and migration, traits that synchronize reproduction with favorable plant resources. Because life history traits of phytophagous insects are influenced in part by spatial and temporal variation in the quality and availability of their host plants, we also consider the role that discontinuities in plant quality play in reducing insect fitness. Lastly, much of the traditional life history theory concerns itself with differences between the evolution of traits or constellations of traits when populations incur primarily density-independent, compared to density-dependent, mortality. Consequently, we address this issue and attempt to shed light on the equilibrium status of several phytophagous insect communities.

Because we consider mostly phytophagous insects here, we hope this work will not only further our knowledge of basic ecological processes, but also stimulate and prove useful to applied scientists attempting to manage pests on agricultural crops, forest trees, and ornamental plants. Furthermore, many of the processes and interactions we discuss are not unique to insects and should extend beyond the bounds of Entomology. Finally, our subject is a broad one that encompasses several controversial issues, and in no way do we perceive this volume as either a complete coverage or the final word on variation in insect life histories. However, we have tried to identify the kind of holistic approach necessary to build and improve upon our currently limited knowledge.

We have divided this volume into three parts. Part I is concerned with the impact of different levels of host plant variability on herbivorous insects. Two chapters discuss the effect of intraplant and interplant variability respectively on insect fitness traits, while the third emphasizes how coevolution with different species of host plants has led to disparate life histories and host race or possibly species formation. Part II includes five chapters that directly discuss insect life history traits. These treat the genetic and environmental variances and covariances and identify the selective pressures associated with single traits or combinations of traits. They also discuss inconsistencies between life history data and current theoretical predictions as well as theoretical shortcomings. In Part III we attempt to define the ecological setting under which the life histories of selected herbivorous insects evolved or at least are maintained. The possibility of insect communities without competition and with nonequilibrium coexistence is considered here, as are patterns and processes of habitat exploitation. The necessity for an open-minded approach should be apparent if we are to identify exceptions to current notions of community organization and make appropriate adjustments to evolving ecological theory.

Robert F. Denno
Hugh Dingle

Contents

List of Contributors

DONALD N. ALSTAD Department of Biology, University of Utah, Salt Lake City, Utah 84112, U.S.A.

WILLIAM S. BLAU Department of Entomology, North Carolina State University, Raleigh, North Carolina 27607, U.S.A.

ROBERT F. DENNO Department of Entomology, University of Maryland, College Park, Maryland 20742, U.S.A.

HUGH DINGLE Department of Zoology, University of Iowa, Iowa City, Iowa 52242, U.S.A.

GEORGE F. EDMUNDS, Jr. Department of Biology, University of Utah, Salt Lake City, Utah 84112, U.S.A.

SHELDON I. GUTTMAN Department of Zoology, Miami University, Oxford, Ohio 45056, U.S.A.

CONRAD A. ISTOCK Department of Biology, University of Rochester, Rochester, New York 14627, U.S.A.

MICHAEL J. RAUPP Department of Entomology, University of Maryland, College Park, Maryland 20742, U.S.A.

RICHARD P. SEIFERT Department of Biological Sciences, George Washington University, Washington, D.C. 20006, U.S.A.

WILLIAM B. SHOWERS USDA-SEA-AR, Corn Insects Research Unit, Ankeny, Iowa 50021, U.S.A.

DANIEL SIMBERLOFF Department of Biological Science, Florida State University, Tallahassee, Florida 32306, U.S.A.

DONALD R. STRONG Department of Biological Science, Florida State University, Tallahassee, Florida 32306, U.S.A.

DOUGLAS W. TALLAMY Department of Entomology, University of Maryland, College Park, Maryland 20742, U.S.A.

THOMAS G. WHITHAM Harold S. Colton Research Center, Museum of Northern Arizona, Flagstaff, Arizona 86001 and Department of Biological Sciences, Northern Arizona University, Flagstaff, Arizona 86011, U.S.A.

THOMAS K. WOOD Department of Entomology and Applied Ecology, University of Delaware, Newark, Delaware 19711, U.S.A.

Insect Life History Patterns: Habitat and Geographic Variation

Introductory Chapter

Considerations for the Development of a More General Life History Theory

ROBERT F. DENNO and HUGH DINGLE

A major concern of modern evolutionary ecology is the nature of adaptations both among different species and among populations within species. Frequently, adaptations consist of complex sets of phenotypic characters that function together and co-vary under selection. One such character complex is the collection of traits that constitutes the life history of a species. Because life history "tactics" (see Stearns 1976) involve schedules of reproduction and survival and therefore directly fitness, much empirical and theoretical effort has been expended on understanding these schedules and the environmental constraints that shape them.

Over the course of the last several years, considerable debate has been associated with the development of life history theory. The central concerns usually lie with the much criticized concept of r- and K-selection proposed by MacArthur and Wilson (1967) and reviewed by others (e.g., Pianka 1970, Gadgil and Solbrig 1972, Stearns 1976). Key life history traits are fecundity, size of young, the age to first and peak reproduction, the interaction of reproductive effort with adult mortality, and the variation in these traits among an individual's progeny (Stearns 1976). In stable environments late maturity, multiple broods, a few large young, parental care and small reproductive efforts should be favored (K-selection), while in fluctuating environments species should be characterized by early maturity, many small young, reduced parental care and large reproductive efforts (r-selection) (Stearns 1976).

An alternative hypothesis (bet-hedging) deals specifically with the consequences of fluctuating mortality schedules. For instance, when juvenile mortality fluctuates due to unpredictable environmental conditions, selection favors organisms with the same constellation of traits predicted under K-selection. If fluctuations in juvenile mortality become more predictable as a result of stable environments, organisms that reproduce quickly but more than once are favored (Murphy 1968). With unpredictable adult mortality, selection in fluctuating and stable environments favors combinations of traits predicted by r- and K-selection respectively (Schaffer 1974).

In his critique Stearns (1977) argues that current life history theory is not yet refined enough to be tested by critical experiments because there are several inadequacies resulting from failure to incorporate important biological complexities. These latter include genetics (most life history traits are influenced by many genes), patterns of ontogeny, design constraints (optimal traits or combinations may not be possible ones), and non-stable age distributions. Other difficulties arise because alternative explantions of life history diversity are not considered or cannot be distinguished and because r and K cannot be reduced to units of common currency. There may also be a problem in selecting an appropriate time scale. Given these limitations, it is hardly surprising that about half the evidence available from an array of cases is not consistent with theoretical predictions (Stearns 1977). To improve upon the testing and re-formulation of theory, Stearns suggests that researchers attempt to adhere to six criteria: (1) constant environment rearing to isolate genetic components of variability, (2) direct measurement of those environmental factors involved to explain differences in reproductive traits, (3) assessment of density-dependent and -independent regulation, (4) measurement of year-to-year variation in mortality schedules, (5) rigorous statistical analysis of observed differences, and (6) adequate measurement of reproductive effort. We agree with Stearns that these criteria are important for the design of experiments and the choice of study organisms, and suggest that there are also additional considerations for the development of a comprehensive life history theory.

We must first ask what elements are missing from the available theoretical edifice and what is needed to provide the framework for restructuring the foundation. These questions were the motivation for organizing this symposium, and, as is obvious from the contributions in this volume, providing satisfactory answers will require input from a variety of biological disciplines. Three considerations seem paramount: the need for a more sophisticated assessment of the genetic structure of life histories, the need for a more detailed analysis of environmental constraints and influences, and the need to include a broader range of traits as life history variables, especially those aspects of physiology and behavior which confer phenotypic flexibility.

We first emphasize the importance of understanding the variation and covariation of life history traits and how they relate to resource tracking. Failure to identify the limits of natural variability in traits upon which selection acts may limit life history predictions because there may be a world of difference between the "optimal" and "best available" trait (Stearns 1976). Central to understanding the limits of variation is genetic analysis. Since life histories are characterized by a complex association of traits, they are likely to be under polygenic control. Consequently, the theory for the analysis of such associations comes from quantitative genetics (Dingle et al. 1981). The important questions revolve around how genes influence the collective function of life history traits and how much variation and covariation is gene-based. Once these are answered, one can approach the problem of the apportionment of variation among genetic and environmental sources. The evidence available (Dingle et al. 1977, 1981, Istock 1978, Chapter 7, Lumme 1978, Lumme and Keränen 1978, Vepsäläinen 1978, Derr 1980, Showers, Chapter 6) suggests that not only does the genetic variance arise from different sources (additivity, dominance, epistasis), but also that the relative contribution of each becomes important to understanding the evolution of complex life histories. The role of environmental uncertainty in maintaining additive genetic variance has been the subject of some recent discussion (Dingle et al. 1977, Hoffman 1978, Istock 1978), and is further considered by Istock in his Chapter 7. Explicit consideration of such gene-environment interactions should lead to new perspectives in the development of further theory (papers in Dingle and Hegmann 1981).

Our second important consideration is the analysis of environmental constraints, especially as they involve the distribution of resources in time and space. Because most of the insects discussed in this volume are phytophagous, we stress in particular the problems that host plants pose for the insects feeding on them and the resulting consequences for life histories. Careful analysis of host plants indicates that variation in plant quality and availability occurs at several levels (within an individual, among individuals, and among species), and that such variability can have profound effects on the evolution of life histories (Whitham, Chapter 1, Edmunds and Alstad, Chapter 2, Wood and Guttman, Chapter 3). Constraints imposed by host plants may result in sufficient divergence of life cycles for speciation to occur (Wood and Guttman, Chapter 3) adding a further dimension to the analysis of life history evolution. Finally, taxa of host plants with well-defined chemical toxins may develop their own associated fauna (Seifert, Chapter 12), allowing analysis of the impact of similar environments (hosts) on dissimilar phylogenies.

On the other hand, similar taxa of both insects and food plants may occur in an assortment of habitats over a wide geographic range. The classic comparison has been between species or populations from temperate and tropical areas, and indeed life histories have been described as fitting these climatic extremes. Temperate-tropical comparisons are central to three papers in this volume (Blau, Chapter 5, Dingle, Chapter 4, Seifert, Chapter 12), and it is apparent that, at least for differences among populations within species, sweeping generalizations concerning life history evolution are no longer tenable. To understand the differences observed, it is necessary to determine the fine structure of habitat variation. Small-scale patterns of temporal and spatial heterogeneity may be as essential to the specifics of life history variation as large-scale climatic differences.

Another significant environmental aspect of the development of life history theory has been the relative contribution of density-dependent and density-independent mortality. Current assumptions concerning competition and the equilibrium status of populations and communities may need reassessment (Price 1980, Denno et al., Chapter 9, Strong, Chapter 10, Simberloff, Chapter 11). If this is indeed the case, those traits assumed to arise in response to competitive interactions (e.g., many characteristics of "K-strategists") may not in fact have done so, and new models will be needed to account for their existence and their covariance relationships with other character complexes. Important to those models will be detailed elucidation of what is happening to the environment in both ecological and evolutionary time and what the appropriate gene-environment interactions are.

Consideration of these interactions leads us to our third point with respect to the needs of a developing life history theory, the inclusion of a broader range of traits. Conventional theory has traditionally addressed those characters associated with reproduction and longevity. Clearly, however, it is as important for a species to place offspring appropriately on a favorable resource as it is to reproduce at all. The ability to track resources in time and space is an important factor in the shaping of life histories (Southwood 1962, Dingle 1972, 1974, 1978, 1979, Southwood et al., 1974, Stearns 1976, 1977, Solbreck 1978, Denno et al. 1980, Price 1980), and the importance of behavioral and physiological flexibility to such tracking is apparent (Nichols et al. 1976, Taylor and Taylor 1977, 1978, Livdahl 1979, Taylor 1981).

In insects, migration and diapause synchronize reproduction with resources in space and time respectively. Consequently we view these behaviors as important elements of life histories; examples of how they integrate reproduction and resources are provided by Kennedy (1975), Solbreck (1978), Vepsäläinen (1978), Denno and Grissell (1979), Denno et al. (1980), Derr (1980) and by Blau, Denno et al. and Dingle in this volume. There is close behavioral and physiological integration of migration and diapause (Kennedy 1961, Dingle 1968, 1972, 1974, Chapter 4, Solbreck 1978), with the strong implication that (energy) trade-offs are involved (Tallamy and Denno, Chapter 8). For instance, energy expended on migration (wings, flight muscles, and associated biochemistry) might be otherwise spent on reproduction; a similar argument applies to the maintenance energy necessary for diapause. Delay in reproduction resulting from diapause may increase the chance of dying before any young are produced, which must be weighed against the potential survival of the offspring themselves. The extent to which trade-offs exist, if they do, will be difficult to measure, not least because of the problem of finding appropriate units. Technical problems, however, should not be allowed to inhibit the development of a comprehensive life history theory which includes relevant behavior.

Migration and diapause are, of course, not the only potentially important behavioral traits. Certain environmental conditions may select for maternal care and other related behaviors such as concealment or defense of young. In terms of life histories, these may be acceptable alternatives to changes in reproductive schedules (Tallamy and Denno, Chapter 8). For example, we have no adequate theory to assess the relative advantages for an insect of inserting eggs one millimeter deeper into a stem (thus conferring greater protection from parasites or severe temperature) versus laying more exposed eggs but increasing fecundity or advancing age at first reproduction. Behaviors such as embedding eggs probably coevolve with more traditional life history traits (Denno et al., Chapter 9), but as yet they have been largely ignored or insufficiently distinguished from design constraints.

The latter should not be confused with trade-offs or costs (Stearns 1977). To illustrate, the lack of an ovipositor is a design constraint, but to have one and spend time and energy inserting eggs at greater depths in vegetation or soil that could otherwise be allocated to fecundity is a trade-off. Again, in order to assess constraints imposed by designs or trade-offs, the question of common units must be carefully considered. Nevertheless, behaviors such as concealment of eggs are important because they confer flexibility to life histories and influence the evolution of coadaptive complexes. They should thus be important constituents of theoretical predictions concerning what collections of traits will occur under given sets of environmental conditions. By influencing the evolution of other traits, fixed features of design (ovipositors, raptorial appendages, defensive spines and the like) can contribute significantly to the disparate life histories of related species that occur in the same habitats (Stearns 1977, Tallamy and Denno, Chapter 8). The need for careful autecological research to find the arena in which life histories are played out should be apparent. Only with such studies can evolutionary ecologists develop truly predictive theory which distinguishes between testable alternatives.

Obviously a volume such as this cannot address in detail all the issues we have adumbrated here. We hope to focus attention on two main issues in studies of life histories.

The first is the need for greater care and greater attention to detail in designing (long-term) experiments and planning fieldwork. This caveat is hardly new and has been discussed in detail by Stearns (1976, 1977) and others. We stress however that, in considering the environment, old assumptions, e.g., concerning host plant constancy or competition, may no longer be valid. The second issue is the need for a revised and expanded perspective incorporating genetic structure (and gene-environment interactions) and a wider array of traits, especially those behaviors which confer flexibility. The papers in this volume are a start toward developing these themes.

References

Denno, R. F., Grissell, E. E.: The adaptiveness of wing-dimorphism in the salt marsh-inhabiting planthopper, *Prokelisia marginata* (Homoptera: Delphacidae). Ecology 60, 221-236 (1979).

Denno, R. F., Raupp, M. J., Tallamy, D. W., Reichelderfer, C. F.: Migration in heterogenous environments: Differences in habitat selection between the wing-forms of the dimorphic planthopper, *Prokelisia marginata* (Homoptera: Delphacidae). Ecology 61, 859-867 (1980).

Derr, J. A.: The nature of variation in life history characters of *Dysdercus bimaculatus* (Heteroptera: Pyrrhocoridae), a colonizing species. Evolution 34, 548-557 (1980).

Dingle, H.: The relation between age and flight activity in the milkweed bug, *Oncopeltus*. J. Exp. Biol. 42, 269-283 (1965).

Dingle, H.: Life history and population consequences of density, photoperiod, and temperature in a migrant insect, the milkweed bug *Oncopeltus*. Am. Nat. 102, 149-163 (1968).

Dingle, H.: Migration strategies of insects. Science 175, 1327-1335 (1972).

Dingle, H.: The experimental analysis of migration and life history strategies in insects. In: Experimental Analysis of Insect Behavior. Barton Browne, L. (ed.). New York: Springer, 1974.

Dingle, H.: Migration and diapause in tropical, temperate, and island milkweed bugs. In: Evolution of Insect Migration and Diapause. Dingle, H. (ed.). New York: Springer-Verlag, 1978.

Dingle, H.: Adaptive variation in the evolution of insect migration. In: Movement of Highly Mobile Insects: Concepts and Methodology in Research. Rabb, R. L., Kennedy, G. G. (eds.). Raleigh, North Carolina: North Carolina State University Press, 1979.

Dingle, H., Brown, C. K., Hegmann, J. P.: The nature of genetic variance influencing photoperiodic diapause in a migrant insect, *Oncopeltus fasciatus*. Am. Nat. 111, 1047-1059 (1977).

Dingle, H., Blau, W. S., Brown, C. K., Hegmann, J. P.: Genetics of life histories: Co-adaptation, covariation, and population differentiation. Nature (1981) (in press).

Dingle, H., Hegmann, J. P. (eds.): Variation in Life Histories: Genetics and Evolutionary Processes. New York: Springer-Verlag, 1981 (in press).

Gadgil, M., Solbrig, O. T.: The concept of r- and K-selection. Evidence from wild flowers and some theoretical considerations. Am. Nat. 106, 14-31 (1972).

Hoffman, R. J.: Environmental uncertainty and evolution of physiological adaptation in *Colias* butterflies. Am. Nat. 112, 999-1015 (1978).

Istock, C. A.: Fitness variation in a natural population. In: Evolution of Insect Migration and Diapause. Dingle, H. (ed.). New York: Springer-Verlag, 1978.

Kennedy, J. S.: A turning point in the study of insect migration. Nature (London) 189, 785-791 (1961).

Kennedy, J. S.: Insect dispersal. In: Insects, Science, and Society. Pimentel, D. (ed.). New York: Academic Press, 1975.

Livdahl, T. P.: Environmental uncertainty and selection for life-cycle delays in opportunistic species. Am. Nat. 113, 835-842 (1979).

Lumme, J.: Phenology and photoperiodic diapause in nothern populations of *Drosophila.* In: Evolution of Insect Migration and Diapuase. Dingle, H. (ed.). New York: Springer-Verlag, 1978.

Lumme, J., Keränen, L.: Photoperiodic diapause in *Drosophila lummei* Hackman is controlled by an X-chromosal factor. Hereditas 89, 261-262 (1978).

MacArthur, R. H., Wilson, E. O.: The Theory of Island Biogeography. Princeton, N.J.: Princeton Univ. Press, 1967.

Murphy, G. I.: Pattern in life history and the environment. Am. Nat. 102, 391-403 (1968).

Nichols, J. D., Conley, W., Batt, B., Tipton, A. R.: Temporally dynamic reproductive strategies and the concept of r- and K-selection. Am. Nat. 110, 995-1005 (1976).

Pianka, E. R.: On "r" and "K" selection. Am. Nat. 104, 592-597 (1970).

Price, P. W.: Evolutionary biology of parasites. Princeton, N.J.: Princeton Univ. Press, 1980.

Schaffer, W. M.: Selection for optimal life histories: the effects of age structure. Ecology 55, 291-303 (1974).

Solbreck, C.: Migration, diapause, and direct development as alternative life histories in a seed bug, *Neacoryphus bicrucis.* In: Evolution of Insect Migration and Diapause. Dingle, H. (ed.). New York: Springer-Verlag, 1978.

Southwood, T. R. E.: Migration of terrestrial arthropods in relation to habitat. Biol. Rev. 37, 171-214 (1962).

Southwood, T. R. E., May, R. M., Hassell, M. P., Conway, G. R.: Ecological strategies and population parameters. Am. Nat. 108, 791-804 (1974).

Stearns, S. C.: Life history tactics: A review of the ideas. Q. Rev. Biol. 51, 3-47 (1976).

Stearns, S. C.: The evolution of life history traits: A critique of the theory and a review of the data. Ann. Rev. Ecol. Syst. 8, 145-171 (1977).

Taylor, F.: Optimal switching to diapause in relation to the onset of winter. Theor. Pop. Biol. (1981) (in press).

Taylor, L. R., Taylor, R. A. J.: Aggregation, migration, and population mechanics. Nature (London) 265, 415-421 (1977).

Taylor, L. R., Taylor, R. A. J.: The dynamics of spatial behavior. In: Population Control by Social Behavior. Ebling, F. J., Stoddart, D. M. (eds.). London: Institute of Biology, 1978.

Vepsäläinen, K.: Wing dimorphism and diapause in *Gerris*: Determination and adaptive significance. In: Evolution of Insect Migration and Diapause. Dingle, H. (ed.). New York: Springer-Verlag, 1978.

PART ONE
Host Plant Variation and Insect Life Histories

Herbivorous insects must track their host plants in both space and time to synchronize reproduction with favorable conditions. Until recently, discussions of host tracking have focused on locating appropriate host species amid an array of unacceptable ones, contending with the defensive and nutritional syndrome of the host, interacting with other herbivores and predators, and existing during periods of time when hosts are absent. The emphasis in studies of plant-insect interactions has been at the species level, with some regard for interplant differences and the effects they might have on herbivore fitness. We have all noticed that certain trees in a "uniform" stand of oaks are galled while others are not, and that differences occur in the size of aphid infestations among conspecific hosts. Observations like these are consistent with a growing block of literature suggesting that intraspecific differences in host plant suitability do exist, and add another dimension to the resource tracking problems of insects. However, a relatively unexplored aspect of the problem is the degree to which an individual plant represents a mosaic of opportunities for herbivores. Are all leaves, buds, twigs and branches equally suitable for development? If not, then the range of host plant variability with which insects must contend should be extended to include these differences. Consequently, the life histories of insects are shaped in part by selective pressures associated with several scales of host plant heterogeneity. In the chapters that follow, the life histories of several herbivorous insects are discussed, each referenced to a different scale of host plant variability.

In Chapter 1, T. G. Whitham addresses within-tree variability in the susceptibility of narrowleaf cottonwood leaves to attack by *Pemphigus* gall aphids. He proposes several mechanisms explaining intratree variability, and discusses the role heterogeneity plays in determining settling and feeding decisions, aphid life history traits, competitive interactions and visibility to predators. In Chapter 2, G. F. Edmunds and D. N. Alstad consider intraspecific defensive variation in pines and how it promotes coevolution of scale insects with individual trees. As a result, however, scale insects are forced into an evolutionary dilemma between colonization of new hosts and adaptation to single ones, consequently placing them at a disadvantage in their evolutionary chase of pines. Both Whitham and Edmunds and Alstad stress the importance of intra- and intertree variability in countering the evolutionary advantage of herbivores with shorter generation times and greater recombination potential than their host plants.

Association of a herbivore with several different species of host plants may select for genetically distinct populations, each with a unique combination of life history traits adapting them to the particular phenology, defensive syndrome and nutrition of their host. Ultimately, coevolution may result in a monophyletic assemblage of reproductively isolated specialists. In Chapter 3, T. K. Wood and S. I. Guttman argue that a complex of treehoppers has evolved in this way, with their life histories fashioned by host plant constraints and mutualistic ants that protect them from predators. They suggest that the timing of reproduction and dispersal on the respective host trees makes an important contribution to life cycle differences maintaining the integrity of treehopper species.

Chapter 1

Individual Trees as Heterogeneous Environments: Adaptation to Herbivory or Epigenetic Noise?

THOMAS G. WHITHAM

1 Introduction

The aim of this paper is to examine the hypothesis that individual host plants are mosaics of varying susceptibilities to parasite and herbivore attack, and that these insects cannot or are not likely to be adapted to all parts of the mosaic at once. Thus, within an individual plant, variation may be an evolved trait that negates the evolutionary advantages of insects which have shorter generation times and greater recombination potential than their host plants. Within-plant variation at all levels (within a leaf, between leaves, between branches) may make the plant appear to be a chameleon of different or changing resistances to herbivore attack. Such variation would pose formidable problems for most herbivores. For example, parasites and herbivores unable to precisely track or discriminate between such variation may either make inappropriate settling decisions and consequently suffer reduced fitness, or, if they discriminate between different plant parts varying in quality, they clump at specific sites on the plant. Due to clumping, herbivores may become more visible to predators, and they may also be forced to engage in competitive interactions for the best resources of the host plant. In each case, the plant benefits by reducing the impact of a debilitating organism. Thus, heterogeneity within a plant may make the plant less apparent to its parasites and herbivores while simultaneously making these same insects more obvious to their predators.

An alternative explanation to within-plant variation being an adaptive response to herbivory, parasitism and *other* factors is that epigenetic or nongenetic factors such as drought, wind, mineral deficiencies, etc., produce variation that is not adaptive, or at least has no selective advantage against herbivory or parasitism.

Variation within a plant may be produced by at least four mechanisms, three of which are genetic: (1) induced factors such as pest-induced plant defenses, (2) somatic mutations and chimeras, (3) developmental patterns of plant growth, and (4) epigenetic factors. I will briefly review these mechanisms and then present empirical evidence for variation within individual trees of narrowleaf cottonwood, *Populus angusti-*

folia, and the resulting effects upon the gall aphid, *Pemphigus betae.* These data support the hypothesis that variation within individual plants may be an adaptive response to herbivory.

2 Mechanisms Producing Within-Plant Variation

2.1 Pest-Induced Plant Defenses

The feeding of herbivores can induce a defensive response in the host plant. Depending upon the extent of the plant response (i.e., whether or not the defensive response is local or spread throughout the entire plant), pest-induced plant defenses may tend to increase the heterogeneity of a single plant. Thus, depending upon past herbivore loads, new colonizers may encounter a mosaic of plant defenses.

Green and Ryan (1972) discovered that damage to a single leaf of either potato or tomato plants by the Colorado potato beetle could induce a defensive response in other non-damaged leaves. Feeding by either adults or larvae, or simple mechanical tearing of a leaf is sufficient to cause the release of a proteinase inhibitor inducing factor (PIIF) into the vascular system of the plant. PIIF is rapidly transported to other leaves where two proteinase inhibitor proteins soon begin to accumulate. Within 48 hours of severely wounding single leaves, the inhibitor proteins can account for 10% of the total soluble proteins in the remaining leaves of the plant (Gustafson and Ryan 1976). Both of these plant proteins attack the digestive proteinases of insect pests and are potent inhibitors of chymotrypsin, trypsin and subtilisin. The accumulation of inhibitor proteins in leaves varies in proportion to the insect damage inflicted upon the plant. Plants of the same age exhibit a 5- to 10-fold variation in the amount of inhibitor. Since PIIF-like activities have now been identified in 20 diverse plant families, this pest-induced plant defense mechanism appears to be widespread (MacFarland and Ryan 1974).

In another example, Haukioja and Niemalä (1976, 1977, 1979) showed that an apparently induced defensive response resulted when leaves of mountain birch, *Betula pubescens,* were mechanically damaged. The induced response was produced in undamaged leaves by mechanically tearing the adjacent leaves on the same shoot. Two days later the undamaged leaves were fed to larvae of the moth, *Oporinia autumnata.* Compared to control larvae fed on leaves from the same tree but not with nearby leaves damaged, the experimental group suffered reduced growth. Similar results were obtained in other experiments with the moth, *Brephos parthenias,* and two species of saw-flies, *Pristiphora* sp. and *Pteronidea* sp. Furthermore, when *Oporinia autumnata* larvae were fed leaves from trees that had been defoliated one or two years previously, a similar negative effect on larval growth resulted, indicating that plant responses may be long-term. It was also determined that fertilized plants in good condition exhibited a stronger negative effect upon female larval growth than plants which had been stressed by having their roots severed.

Although examples of plant defenses induced by insects and vertebrates are limited, there are numerous examples of fungus-induced plant defenses (Keen et al. 1971,

Matta 1971). In fact, a whole class of compounds, the phytoalexins, consists of inhibitory substances that accumulate in host tissues in response to invasion by fungal parasites (Kuć 1972). Müller (1953) showed that, when a strain of potatoes resistant to late blight was inoculated with a virulent race of *Phytophthora infestans,* phytoalexins were produced, inhibiting the development of this virulent race.

2.2 Somatic Mutations

Whitham and Slobodchikoff (1981) developed the hypothesis that somatic mutations may be an important source of heritable variation in plants. Due to shorter generation times and greater recombination potential, insect herbivores should be able to break the defenses of long-lived plants. Somatic mutations might allow plants to compensate for the apparent advantage of herbivores. If this is the case, individual plants may change in gene frequency through time and/or develop as mosaics of genetic variability. Two factors make this a reasonable hypothesis: First, an individual plant may more appropriately be considered as a population (Harper 1977) because a single plant or clone is generally composed of thousands of buds, each of which is self-replicating. Second, somatic mutations in plants are heritable and can be preserved by naturally occurring sexual and asexual mechanisms of reproduction (Dermen 1960, Neilson-Jones 1969, Grant 1975, Hartmann and Kester 1975, Stewart and Dermen 1979).

Several mechanisms exist that make it possible for an individual plant to develop as a genetic mosaic and these are not dependent upon somatic mutation rates. For example, mutations often occur in which only a portion of the meristem is affected, and the resultant plant is composed of two or more genetically distinct tissues growing adjacent to one another. These plants are termed "chimeras." With sectorial chimeras, the genetic composition of future buds is dependent upon their position of origin on the stem. Buds arising on one side will be of one genotype, buds arising on the opposite side will be of the other genotype, and buds arising at the juncture of both genotypes will carry both genotypes in varying proportions (Neilson-Jones 1969). Chimeras of *Pelargonium* exhibit this kind of variation and with sexual reproduction, the variation is passed to seedlings (Baur 1930). While self-pollinated flowers on totally green branches give rise to only green seedlings and self-pollinated flowers on totally colorless branches give rise to only colorless seedlings, a flower on a green branch that is cross-pollinated with a flower from a colorless branch will produce seedlings that develop into green, colorless and chimeral branches. Other mechanisms that may increase variation above the level expected from normal somatic mutation rates include genetic "hot spots" that mutate with high frequency to alternative stable states, and adventitious buds, which may express an alternative genotype when plants are stressed.

Because only those somatic mutations affecting color and growth form are easily noticed, most probably go undetected. Even so, somatic mutations have been a major source of commercially important fruits and ornamental trees. Pink-fleshed grapefruit were discovered in 1906 when the grapefruit from only one branch of one tree was found to possess pink flesh (Hartmann and Kester 1975). Before the use of mutagenic chemicals, 33% of the plant patents issued by the U.S. Patent Office were for somatic mutants (Shamel and Pomeroy 1936). Furthermore, of the 8000 varieties of plants cultivated in Europe in 1899, 5000 originated as somatic mutants (Cramer 1907, cited in Shamel and Pomeroy 1936).

Long lifespan, the complete regeneration of buds each year, and large clone size increase the probability that somatic mutations will arise and that an individual plant will become a mosaic of different gene frequencies. For example, individual clones of bracken fern, *Pteridium aquilinum,* may be 1,400 years old and cover an area of 138,400 m^2 (Oinonen 1967a, 1967b). Similarly, clones of goldenrod, *Solidago missouriensis,* may be 1,000 years old and contain up to 10,000 stalks (William J. Platt, pers. comm.). Vasek (1980) has estimated that some clones of creosote bush, *Larrea divaricata,* are more than 10,000 years old. Considering the large numbers of buds involved over such long periods of time, it seems likely that these plants may have changed in their gene frequencies and/or developed as mosaics.

2.3 Developmental Patterns

Due to developmental factors alone, individual plants can become a highly heterogeneous habitat for herbivores. Developmental factors are largely under genetic control and may affect both the distribution of plant defenses and the resources that herbivores require. Here I consider maturation or the change from juvenile to adult plant growth and the superimposed patterns of apical dominance.

Within the same plant, both juvenile and adult phases can occur simultaneously and both may vary greatly in traits that affect herbivores. When the upper branches of a tree reach the adult phase, the lower branches often remain in the juvenile condition. Differences between adult and juvenile branches include growth rate., flowering capacity, leaf and stem anatomy, and thorniness (Kozlowski 1971, Zimmerman and Brown 1971). Honeylocust, *Gleditsia triacanthos,* constitutes an example of this type of situation. The base of the honeylocust tree has numerous large thorns that tend to deter large herbivores from damaging these thin-barked trees. The upper branches of the tree, however, do not possess thorns. The thorny condition represents a juvenile trait, while the thornless condition represents an adult trait. O'Rourke (1949) demonstrated that the loss of thorns by adult tissues is permanent. Rooted cuttings collected from the top of the tree retained the adult characteristics and failed to develop thorns, while rooted cuttings collected from the base of the tree developed normal thorns.

Superimposed upon the patterns of maturation are patterns of apical dominance that concentrate the resources of the plant at specific sites, resulting in differences between adjacent leaves and shoots. Apical dominance is the result of auxins produced by the actively growing apex or terminal buds that inhibit the growth of lateral buds beneath. Kozlowski and Ward (1961) found that the terminal leaders of *Pinus resinosa* exhibited as much as 12 times more growth than lateral buds. Terminal buds are also more likely than lateral buds to continue producing leaves beyond the initial flush of spring growth (Kozlowski 1971). Thus, some shoots contain both mature and developing leaves, while adjacent shoots may contain only mature leaves.

The above patterns of growth within a single plant (juvenile vs. adult growth and apical dominance) are very important to herbivores because they produce resource sinks that vary in time and space. Woodwell et al. (1975) showed that the concentrations of nitrogen, phosphorus, potassium, calcium, magnesium, sulfur, iron and sodium varied greatly in different parts of the same plant and were concentrated in the most metabolically active tissues. Because the same shoot may contain leaves of different ages produced during different spurts of growth, even adjacent leaves on the same shoot may differ greatly in their suitability to herbivores. Many studies of different plant species have shown that as leaves expand, mature and senesce, striking changes occur in the rate of photosynthesis and the accumulation of chlorophyll, nucleic acids and proteins (Spenser and Titus 1973, Woodward and Rawson 1976, Lurie et al. 1979).

Also associated with leaf age are patterns of digestibility-reducing substances and toxins. Feeny (1970, 1976) compared the growth of the winter moth, *Operophtera brumata,* on developing and mature leaves of oak (*Quercus robur*). The caterpillars did best on the developing leaves which were high in nitrogen and water content. Mature leaves were tough, lower in nitrogen and water, and high in tannins that can combine with proteins and inhibit digestion. Rhoades and Cates (1976) showed that developing leaves contained higher concentrations of various toxins (alkaloids, cyanogenic glycosides, cyanide, mustard oils, and others) than mature leaves, whereas mature leaves tended to have more digestibility-reducing substances (tannins). Thus, in plants with leaves at various stages of development, herbivores are likely to encounter different plant chemical defenses.

2.4 Epigenetic Factors

An alternative explanation for within-plant variation being an adaptation to herbivory or other selective factors is that such variation is simply the result of non-genetic or environmental factors. Environmental factors may not only produce variation within a single plant: they may also interact with genetic factors.

Because the extensive root system of a tree encounters different soils, nutrients, water levels, etc., the branches that a single root services may reflect the environment the roots are encountering. With *Quercus* and other ring porous species, water ascends from one region of the root system to one particular portion of the crown (Zimmermann and Brown 1971). Another example of branches being independent was observed with elm trees, following salt-water flooding of fresh-water drainage ditches in Holland. Damage was restricted to that side of the tree facing the ditches, indicating that specific flow patterns exist between the affected branches and the roots drawing water from the drainage ditch (Richardson 1958).

Environmental stress in the form of nutrient deficiencies affects parts of the same plant very differently (Kozlowski 1971). Deficiencies in copper, zinc, iron and calcium have a much greater effect on young leaves than on mature leaves by causing leaf chlorosis. Conversely, deficiencies of potassium, phosphorus and magnesium have a greater impact on mature foliage where these nutrients are mobilized and translocated to young leaf tissues. Similarly, deficiencies of these nutrients, with the exception of zinc, affect parts of the leaf blade differently. The base of the leaf blade and midrib are least affected, with the greatest chlorosis or discoloration occurring at the leaf tip and margins. Even different branches of the same tree may be affected differently. Nitrogen deficiencies result in translocation of nitrogen from the oldest shoots and branches to the most actively developing meristems. Magnesium deficiencies are most noticeable in the upper crown of the tree.

It is important to point out that the factors which produce variation within plants may interact with one another. For example, the nonsynchronous behavior of the crown of some tropical trees may be due to the interaction of genetic and epigenetic factors. With *Tabebuia serratifolia,* adjacent branches may be in very different developmental stages. Some may be flowering and not possess foliage, others may be leafing out, and still others may contain mature foliage. Hallé et al. (1978) suggest that such nonsynchrony is related to the formation of adventitious buds which express the juvenile condition and cause different plant parts to be out of phase. Adventitious buds are often formed when existing buds are damaged by environmental factors or herbivory. In addition, the environments of different roots may enhance this out-of-phase condition.

Since the top of the tree is under the greatest stress due to the forces required to lift water against pressure gradients that increase with height, water stress affects parts of the same plant differently. Such environmentally caused stress may strongly interact with the genetic systems of the plant. Studies of both herbaceous and woody plants show that the upper leaves of the crown are more xeromorphic than leaves lower on the tree. Similarly, sun leaves, especially those on the south side of trees in north temperate regions, are more xeromorphic than shade leaves from the interior of the crown. These differences become even greater with increased moisture stress (Shields 1950, Zimmerman and Brown 1971). The heavy cuticle, thick-walled cells and increased mechanical tissues of xeromorphic leaves are undoubtedly important to herbivores and should affect their ranking of potential food items. Water stress is also strongly correlated with insect outbreaks. It has been proposed that such outbreaks are due to an increase in the amount of nitrogenous food that becomes available to herbivores when plants are physiologically stressed (White 1969, 1978). Since nitrogenous compounds are not distributed equally throughout the plant, stress is likely to increase the level of nitrogen in some plant parts more than in others. There may also be a reduction in the chemical defense system of the plant that accompanies stress. Rhoades and Cates (1976) and Feeny (1976) showed that chemical defenses are not distributed equally throughout the plant, and suggest that digestion-inhibiting substances are much more costly to produce than toxins. Consequently, stress may significantly reduce the defenses in one part of the plant while having less effect on other plant parts.

3 An Empirical Example of the Consequences of Host Variation

3.1 Levels of Variation within a Single Host Plant

For the gall aphid, *Pemphigus betae,* individual host plant parts of the same narrowleaf cottonwood tree, *Populus angustifolia,* are highly variable in suitability. This variation affects the survival of colonizing stem mothers during gall formation, the number of progeny of the survivors, the development rate of progeny to maturity, the weight of the progeny, and the number of embryos contained in mature progeny. As a result, colonizing stem mothers have evolved discriminatory behavior which allows them to preferentially settle at superior gall sites. However, these superior gall sites are few in number and may become limiting even at relatively low densities of colonizing aphids. The remainder of this paper will examine the consequences of variation within a single host plant to determine whether or not such variation may be adaptive.

In early spring when leaf buds first begin to break, colonizing stem mothers of *P. betae* emerge from overwintering eggs in the fissures of the tree trunk and migrate to the developing leaves. Within four or five days of bud break, most colonizing stem mothers have permanently settled, and gall formation has been initiated. Using their stylets to methodically probe immature leaf tissues to induce gall formation, each stem mother is soon enclosed within a rapidly expanding hollow gall, which at maturity measures approximately 1.5 cm in length by 0.6 cm in width. Within this gall, up to 300 progeny are produced parthenogenetically. Since each stem mother produces either a successful gall or dies during gall formation, leaving a scar as evidence of the attempt (Whitham 1978), a census of successful and aborted galls accurately reflects the colonizing population. By counting the number of progeny within mature galls, the reproductive success of surviving stem mothers can be measured. With these data, one can quantitively examine the patterns and effects of host variation on aphid selectivity during colonization and the resultant reproductive success of individuals.

At least three levels of variation have been quantified within a single host tree. Variation within a single leaf, between leaves of the same branch, and between adjacent branches of the same tree all affect aphid survival and reproduction. Although the precise mechanism(s) of variation within an individual tree is as yet unclear, it does have a biochemical basis. Zucker (1981) has recently established a strong negative correlation relating individual reproductive success to the concentration of secondary compounds.

Starting with the smallest level of variation presently known, variation within a leaf, there exists a key position for successful gall formation on a single leaf (Figure 1-1). Whitham (1978) showed that, when several galls of *P. betae* occupied the same leaf blade, the stem mother in the basal position achieved much greater measures of relative fitness. With three stem mothers or galls arranged linearly along the midrib of a single leaf blade, the basal stem mothers were 18 times less likely to die in the gall-forming attempt than the most distal stem mothers. In addition, the surviving stem mothers were 34% heavier in their mature dry body weights than the most distal stem mothers, and they produced 90% more progeny. The stem mothers in the inter-

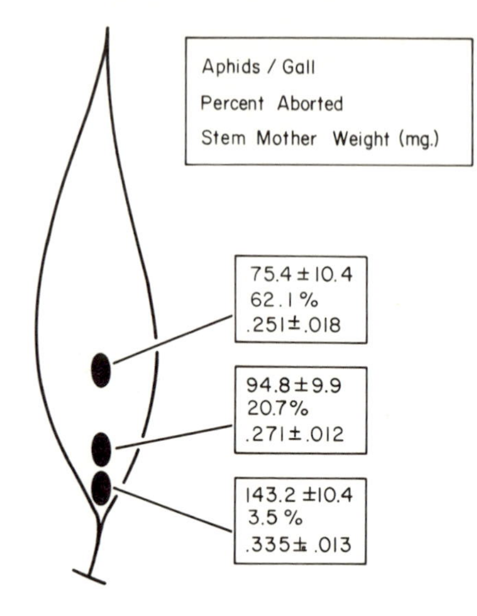

Figure 1-1. Effect of gall position on measures of relative fitness when three *Pemphigus betae* stem mothers colonize the same leaf blade (N = 35 leaves). Shown are the mean number of aphids per surviving gall ± 1 SE. Probability of stem mother's death is greatly reduced at the base of the leaf ($\chi^2 = 25.708$, df = 2, $P \leqslant 0.001$). Analysis of variance (Student-Newman-Keuls test) shows means of basal galls to differ significantly from distal galls ($P < 0.01$). (Adapted from Whitham 1978.)

mediate position achieved intermediate success. The above differences in relative fitness probably represent hard selection, and greatly favor those stem mothers which discriminate in the positioning of their galls. As a result of these selective factors, in the absence of the competitive interactions, stem mothers occupy leaves singly and settle in the optimal position at the base of the leaf blade.

The next level of variation affecting *Pemphigus* aphids is the difference in leaf size between leaves of the same branch. At least six measures of reproductive success are significantly correlated with the size of the mature leaf (Whitham 1978). For example, on the largest leaves available (> 15 cm^2), the probability of stem mothers dying during gall formation was 0% compared to an 80% mortality rate on small leaves ($\leqslant$ 5 cm^2). Similarly, on the largest leaves, the dry body weight of surviving mature stem mothers was 70% greater, and the number of progeny 220% greater, than the progeny of stem mothers occupying small leaves. Clearly, the selective advantages of discriminating between leaves are great, and stem mothers do discriminate with considerable accuracy (Whitham 1980). It is important to note that the distribution of leaf sizes available for colonizing aphids to choose from is highly skewed towards the smaller, less suitable leaf size categories (Figure 1-2). For example, leaves $\leqslant$ 5 cm^2 accounted for about 32% of all available leaves, while the most suitable large leaves (> 15 cm^2) represented only 1.6% of the total leaves available. In spite of the fact that the most suitable large leaves were rare, colonizing stem mothers selectively sought out and colonized all of these leaves, with the average leaf supporting 1.6 galls. However, the numerous and unsuitable small leaves were avoided, with only 3% being

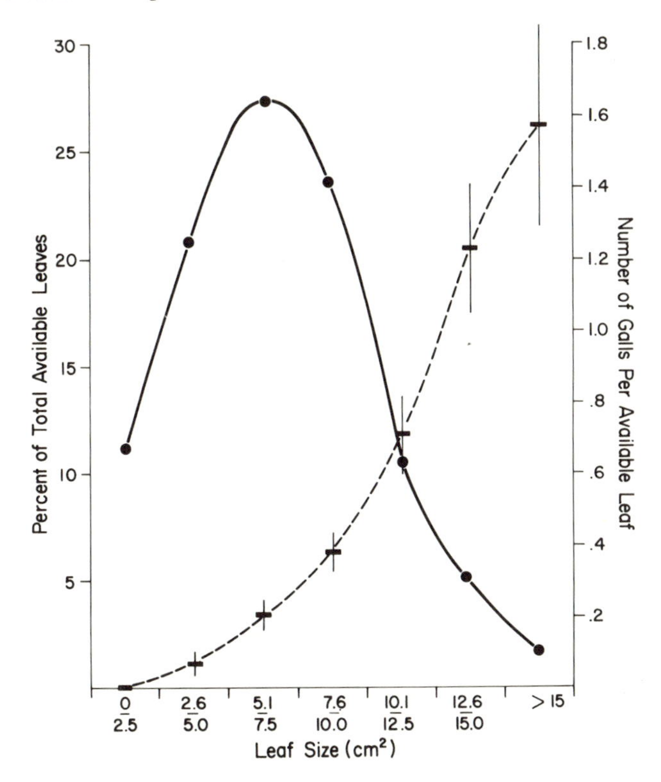

Figure 1-2. Solid circles = the distribution of leaf sizes available for colonization (mean leaf size = 6.9 ± 0.17 cm^2, N = 422); solid bars and vertical lines = means $\pm$ 1 SE of the number of galls per available leaf of each leaf size category (mean leaf size colonized = 10.9 ± 0.27 cm^2 N = 153). At a density of 35 stem mothers per 100 leaves, the mean leaf size colonized is 60% larger than the mean leaf size available. (From Whitham 1980, by permission of The University of Chicago Press.)

colonized. The resultant differential in aphid selectivity is that leaves > 15 cm^2 were occupied at a level approximately 50 times greater than leaves $\leqslant 5$ cm^2.

The third level of variation which has a pronounced effect upon stem mother selectivity and reproductive success is the variation between adjacent branches of the same tree. Table 1-1 shows results from the analysis of leaves randomly sampled from three adjacent branches of the same 22-meter tree (Whitham 1981). These branches occupied different quadrates of the tree trunk 3 meters above the ground. All three branches differed significantly in their mean leaf sizes available for aphid colonization (ANOVA-Student-Newman-Keuls test, $P < 0.001$ for all means), and on each branch only the largest leaves were colonized. On each branch, stem mothers selectively settled on leaves that were approximately 50% larger than the mean leaf size available. Such selectivity within a single branch is also reflected in the discrimination between adjacent branches. The density of stem mothers on the smallest-leaved branch was only 18 per 1,000 leaves, compared to 478 per 1,000 leaves on the largest-leaved branch. This represents a 27-fold difference in aphid selectivity. An evolutionary reason for such differential colonization is the corresponding differences in survival and reproduction. Only 37% of those stem mothers that settled on the smallest-leaved branch survived to reproduce, while 72% survived on the largest-leaved branch ($\chi^2 = 22.988, P < 0.001$). Similarly, the number of progeny per surviving stem mother is nearly twice as great on the largest-leaved branch ($t = 7.361, P < 0.001$), where the average surviving stem mother produced 134 progeny compared to only 75 progeny on the intermediate leaf-sized branch. So few stem mothers survived on the smallest-leaved branch that the sample size was insufficient for analysis. In other comparisons, the branch with an intermediate mean leaf size was also intermediate in stem mother selectivity and survival. Clearly, differences between branches can be great, and those stem mothers that do not discriminate are selected against in terms of reduced survival and reproductive success.

Although this paper is primarily concerned with the variation in host suitability within a single plant, it is important to point out that the variation within a single plant is nearly as great as the variation between trees at the same study site. Table 1-2

Table 1-1. Variation within individual host plants of *Populus angustifolia* affects aphid settling and predator foraging behavior. Shown are data collected from three adjacent branches of the same tree. (Adapted from Whitham 1981.)

	Branch 1	Branch 2	Branch 3
Tree			
Mean leaf size available (cm^2)[a]	5.0 (±0.23, 172)	8.0 (±0.14, 693)	9.9 (±0.18, 904)
Aphid Discrimination and Survival			
Mean leaf size colonized (cm^2)[a]	7.5 (±0.31, 32)	11.7 (±0.26, 162)	14.5 (±0.20, 439)
Density (galls per 1,000 leaves)[a]	17.5 (±3.00, 1886)	24.2 (±2.30, 4630)	477.9 (±21.90, 1333)
Percent survival during coloni-zation[b]	36.7 (33)	61.8 (165)	72.1 (487)
Mean aphids per surviving preda-tor-free gall[a]	–	75.3 (±3.93, 76)	133.6 (±4.71, 194)
Predation			
Percent galls preyed on[b,c]	–	25.4 (102)	44.7 (351)
Effect of Predation on Reproduction			
Mean aphids per gall[a,d]	–	64.8 (±3.84, 102)	85.8 (±4.16, 351)

[a] Numbers in parentheses adjacent to means = ±1 SE, N.
[b] Numbers adjacent to percents = N.
[c] Aborted galls (stem mothers died during colonization) excluded.
[d] Includes predator-free and predator-infested galls (aborted galls excluded).

shows aphid survival and selectivity on five different mature host trees. A 3-fold difference existed in mean leaf sizes between these five trees. Note that on the tree where survival was greatest (87%) the density of colonizing aphids was nearly 24 times greater than on the tree where survival was lowest (24%). Nearly identical relationships have also been shown for two other gall aphid species, *Pemphigus populicaulis* and *Pemphigus populitransversus*, on Eastern cottonwood (*Populus deltoides*) (Whitham, unpublished data).

These data demonstrate that, regardless of the mechanisms producing variation within and between potential host trees, as far as colonizing aphids are concerned, the individual host plant is a mosaic of varying susceptibilities to attack at all levels of host selection (within a leaf, between leaves, between branches, and between trees). Although the ability of aphids to discriminate in their host-selection behavior is adaptive, two questions must be addressed to determine whether or not the tree benefits from the observed variation: First, what levels of aphid infestation constitute sufficient damage that counter-adaptations by the host plant will be evolutionarily favored? Second, if there were no variation within or between host plants, would the aphids or host plants fare any differently? If aphids are not capable of inflicting any real damage on their host plants or if no real benefit is derived from host variation, then heterogeneity can be eliminated as a potential plant adaptation.

3.2 Aphid Impact on the Host Plant

While the impact of aphids upon vegetable crops is well known, less is understood about their effects on long-lived perennials. Several examples demonstrate that the energy and nutrient drain on host trees can be severe. In examining the width of the annual growth ring as a function of the standing crop of aphids, Dixon (1971a) estimated that, in the absence of the sycamore aphid, *Drepanosiphum platanoides*, sycamore trees (*Acer pseudoplatanus*) could produce as much as 280% more stem wood each year. The decline in the production of stem wood is also accompanied by a reduction in leaf area, shoot and root production. Dixon (1971b) has also shown that

Table 1-2. Between-tree variation in suitability for *Pemphigus betae*. Shown are densities and survival rates of colonizing stem mothers on different host trees of narrowleaf cottonwood (*Populus angustifolia*)

	Pemphigus betae	
Tree (mean leaf size, cm^2)[a]	Galls/1000 leaves[b]	Percent survival[c]
4.7 (±0.09, 413)	7.7 (4284)	24.2 (33)
5.1 (±0.11, 415)	0 (1276)	–
6.1 (±0.13, 431)	13.8 (3852)	66.0 (53)
8.8 (±0.14, 546)	105.7 (1542)	78.5 (163)
15.9 (±0.25, 431)	183.8 (1752)	87.0 (322)

[a]Numbers in parentheses = ±1 SE, N.
[b]Numbers in parentheses = No. of leaves examined.
[c]Numbers in parentheses = No. of galls examined.

a 14-meter lime tree, *Tilia x vulgaris,* with 58,000 leaves may carry 1,070,000 lime aphids (*Eucallipterus tiliae*), at one time. The roots of infested lime saplings exhibited little or no growth, and these plants achieved only 7-8% of the weight increase of aphid-free plants. Another example by Fedde (1973) demonstrates that aphids can negatively affect the reproductive capacity of their host plants. Since being introduced into North America around 1900, the balsam woolly aphid, *Adelges piceae,* has caused extensive damage and mortality to stands of Fraser fir (*Abies fraseri*). In comparison with aphid-free trees, infested trees produced fewer and smaller cones, and the seeds were lighter, less viable, and more frequently attacked by the chalcid seed parasite, *Megastigmus specularis* (Table 1-3). The percent germination of seeds from aphid-free trees was 75% compared to only 32% on infested trees. It is particular interesting that the seeds of aphid-infested trees were 10 times more likely to be destroyed by chalcid seed parasites. Consequently, infestation by aphids is not only debilitating in itself, but it also makes the host plant more susceptible to other parasites as well.

It appears that relatively low levels of aphid infestation can negatively affect the host plant. Dixon (1973) showed that the more aphids there are infesting a tree, the greater the reduction in growth. Most importantly, the relationship between the growth of lime and the number of aphids infesting the tree was negatively correlated, and the relationship was linear over a moderate range of aphid densities (5-25 aphids per 100 cm^2 leaf area). Thus, even at relatively low aphid densities, a negative effect on the host plant may be expected. Mittler (1958) calculated that a single willow aphid, *Tuberolachnus salignus,* could in one day ingest the photosynthate of 5-20 cm^2 of leaf. Using energy budget studies of both the aphid and host plant, Llewellyn (1972, 1975) concluded that the observed density of lime aphids in the field (5 per leaf) resulted in an energy drain of 19% of the tree's annual net production. He also calculated that 26 aphids per leaf would consume 100% of the tree's annual net production. These calculations are equivalent to the net production of 1.5 cm^2 of leaf per aphid. Combining Mittler's (1958) and Llewellyn's (1975) estimates, a single aphid could ingest the photosynthate of 1.5-20 cm^2 of leaf.

Using the above estimates of aphid impact upon the host plant, the effect of *Pemphigus betae* gall aphids on narrowleaf cottonwood, *Populus angustifolia,* can be approximated. Using the most conservative ingestion rate (1.5 cm^2 of leaf per aphid), with 5.8 galls per 100 leaves, all of the tree's net production should be lost to aphids. These calculations are based on the mean leaf size of narrowleaf cottonwood trees (about 7 cm^2), a mean of 80 aphids per gall, which includes stem mothers that died (approx. 30%) as having zero progeny, and the assumption that a gall on one leaf

Table 1-3. Comparison of cones and seeds from aphid-free and infested trees of Fraser fir. (From Fedde 1973.)

	Cone Length (mm)	Seed Weight (g)	% Empty Seed	% Seed Parasitized by chalcids	% Germination
Infested	43.5	0.516	42.4	30.9	32.5
Aphid-free	59.7	0.844	28.0	3.1	75.5

$P < 0.01$ in all pair-wise comparisons.

can act as a sink by drawing resources from adjacent leaves (Miles 1968, Way 1968, Way and Cammell 1970, Hori 1974). At the very least, each galled leaf is lost to the tree, because by early June they have become chlorotic and have started to fall from the tree at least two months prematurely. Even if the error in these calculations is substantial, the observed densities of gall aphids in the field are commonly as great or greater (3-50 galls per 100 leaves) than the calculated density levels at which the tree should be damaged. In addition, under natural field conditions in an outbreak year, one can easily find individual trees that are being colonized by 80,000 stem mothers. Including effects of mortality during gall formation and predation, the surviving stem mothers can collectively produce 4.5 million progeny (Whitham 1981). At these densities, aphids should certainly represent a significant drain on the host plant.

3.3 Impact of Host Variation on Aphids

Since the host plant(s) represents a mosaic of varying susceptibilities to attack, colonizing aphids are faced with important settling decisions which affect their survival and subsequent reproductive success. Heterogeneity of the host plant can reduce the impact of aphids in three basic ways: First, heterogeneity favors the evolution of territorial behavior whereby aphids compete for superior gall sites. Second, heterogeneity makes the plant less apparent to aphids and increases the frequency of settling mistakes. Third, heterogeneity tends to increase the clumping of aphids, which makes them more apparent to predators.

High variation in leaf suitability and a leaf size distribution skewed towards the most unfavorable leaves may be mechanisms by which host plants manipulate the competitive interactions of parasites and herbivores to reduce their impact. Once the density of colonizing stem mothers exceeds the available numbers of the most suitable but relatively rare large leaves, competitive interactions result which reduce aphid survival and reproductive success. Whitham (1979) demonstrated that, when the density of colonizing stem mothers exceeds the number of superior large leaves, intense territorial conflicts lasting as long as two days occurred. Large dominant stem mothers force subordinates to adopt one of several strategies: (1) to settle on inferior small leaves; (2) to settle at inferior distal leaf positions; (3) to form a floater population. Since there are only a few days during which a successful gall can be formed in the expanding leaf tissues, any behavior that delays settling has a negative effect on survival. For example, in two replicate experiments using *Pemphigus betae* on *Populus angustifolia* (Whitham 1979) and *Pemphigus populivenae* on *Populus fremontii* (Whitham 1978), it was shown in both cases that the survival rates of stem mothers that successfully defended their gall sites and did not move were nearly three times greater than those of the late settling colonizers from the floater population (61-72% vs. 22-24%, respectively). In addition, stem mothers may spend up to 50% of their time during the colonization phase in the active defense of their gall sites. Such defense may even be at the expense of gall formation because the resultant galls are smaller and contain fewer progeny (Whitham 1979; 1980, unpublished data). The negative effects of colonizing either small leaves or occupying the distal positions of a single leaf have already been discussed in this paper. It would appear that, for the same parasite load, individual plants which encourage competitive interactions between their parasites (high variation in resource quality and relatively rare superior gall sites) would be less affected by parasites than plants that did not encourage competitive interactions (low variation in resource quality or all gall sites equally good).

A second potential source of mortality arising from host heterogeneity is the inability of colonizing parasites and herbivores to make precise habitat selection decisions on a global scale. The basic theory of habitat selection (Fretwell and Lucas 1970, Fretwell 1972) predicts that the decision of whether or not to settle at a specific site should be based upon the expected fitness that would be achieved at that site. Theory predicts that colonizers will adjust their densities in habitats of varying quality in response to their expected fitness within a particular habitat. Whitham (1980) examined these predictions and found that, on a local scale (within a single branch), colonizing stem mothers were very precise in their settling behavior. Expected fitness appeared to be the basis for settling decisions. For example, as the number of galls per leaf increases, the average number of progeny per stem mother declines. However, large leaves can successfully support more galls than small leaves. The data show that colonizing stem mothers adjust their densities on leaves of different size, indicating that they effectively assess and integrate the variables of individual leaf quality, the distribution of leaf qualities available for colonization, and the density of potential competitors. In comparisons between the expected reproductive success that would have been achieved had each stem mother responded perfectly to these variables and the actual reproductive success realized in the field, on a local scale the field population achieved at least 84% of their potential reproductive success.

Although colonizing aphids are quite precise in their habitat selection behavior on a local scale, on a more global scale (e.g., between branches or between trees) the precision of habitat selection is likely to decline. With greater variation or habitat heterogeneity, colonizing aphids are less likely to settle appropriately in terms of their expected fitness. The net effect of settling mistakes should be to reduce the impact of the aphid population on the host plant. For example, Table 1-1 shows that most colonizing aphids avoided small leaved branches within the same tree, but that the aphids actually colonizing those branches suffered far greater mortality, with the few survivors realizing reduced reproductive success. Similarly, Table 1-2 showed that, even though most aphids avoided specific trees, the stem mothers that did colonize those trees suffered much greater mortality. A more critical examination of habitat selection on a global scale is now needed to determine whether the mosaic pattern of within- and between-tree variation makes long-lived perennials less apparent to their parasites and herbivores by increasing the incidence of settling mistakes. This should be particularly true for insects with only a limited time span during which to make a settling decision, after which they are irreversibly committed to a specific leaf, branch or tree (e.g., gall insects). For the same parasite load, it would appear that individual plants which encourage settling mistakes on the part of the parasites and herbivores (high variation in resource quality and low herbivore sampling time) would be less affected by these insects than plants that did not encourage settling mistakes.

A third consequence of within- and between-plant heterogenity is the tendency of plant parasites and herbivores to clump at the most favorable feeding sites which, in turn, makes them more apparent to predators. For example, Gibb (1958, 1962) found that the coal tit selectively foraged on pine trees that had high concentrations of the overwintering eucosmid moth, *Ernarmonia conicolana.* Their foraging was so selective that, by the end of the winter, the number of surviving moth larvae on trees that had an initially heavy moth load was only slightly greater than those trees having an initially low moth load. In this case, clumping increased the probability of an individual larva being preyed upon. With *Pemphigus* gall aphids, host variation and the discriminating behavior of colonizing stem mothers result in the clumping of galls on specific leaves, branches and trees. If colonizing stem mothers settle solely in response to host variation without due regard to potential predators, such behavior may not be adaptive. Within a single branch, Whitham (1978, 1980) found that the various aphid predators did not selectively forage on large leaves nor on multiply galled leaves where aphids were most concentrated. In progressing to a more global scale, however, aphid, predators do discriminate by selectively foraging on those branches where aphid densities are greatest (Whitham 1981). Table 1-1 shows that the high gall densities on preferred branch 3 attracted nearly twice as many predators as branch 2, which was avoided by aphids (45% vs. 25% predation rates respectively; $\chi^2 = 12.149, P < 0.001$). Such predator selectivity places the aphids squarely between two opposing forces. Although differences between branches select for aphid discriminatory behavior and result in aphids clumping on specific branches, selective foraging by predators nearly negates the advantages of colonizing one branch over another. On the preferred branch (3), the average reproductive success of stem mothers that survived colonization (aborted galls excluded) declined from 134 progeny per predator-free gall to 86 progeny per gall when the effects of predators were included. On the branch aphids avoided (2), lower predation levels reduced the net effect. Reproductive success declined from 75 progeny per predator-free gall to 65 progeny per gall when the effects of predators were included. Although branch 3 still remains the most profitable branch for aphids to occupy ($t = 2.627, P < 0.01$), due to the selective foraging of predators, the advantages are greatly diminished. Again, it would appear that the host plant has benefited from variation that clumps herbivores and parasites, thereby making them more vulnerable to predators. Whether or not aphids can appropriately settle on a more global scale in response to both host variation and predators remains to be demonstrated.

4 Concluding Remarks

This paper has examined some of the genetic and epigenetic factors which produce variation within individual host plants. It has also demonstrated how such variation in single cottonwood trees affects the settling behavior and reproductive success of *Pemphigus* gall aphids. From these observations, I suggest that variation may constitute an adaptive plant response to parasites and herbivores because it can negatively affect the pest population in the following ways:

(1) Heterogeneity makes the plant less apparent to its pests by increasing the probability that inappropriate settling and feeding decisions will be made, thereby reducing insect fitness.

(2) Heterogeneity increases the level of competitive interactions for superior host resources. These competitive encounters result in aggressive neglect, floater populations, and other insect behaviors which reduce the cumulative impact of the pests upon the plant.

(3) Heterogeneity results in the clumping of parasites and herbivores on the most favorable host resources, thereby making them more apparent and vulnerable to their predators.

Ecologists generally try to minimize variation and deal with the mean individual or mean population. However, in the case of plants, evolution may have favored a high level of variation within individuals. If this is true, within-plant variation may represent an ecologically more meaningful statistic than the norm. The fact that the variation in host suitability within a single cottonwood tree is nearly as great as the variation between trees suggests that heterogeneity may be more than the result of epigenetic or environmental factors. Although we do not yet know the exact mechanism(s) accounting for the documented variation in the system involving *Pemphigus betae* on *Populus angustifolia,* colonizing aphids respond to individual host plants as if they were a mosaic of varying susceptibilities to successful colonization. The aphids settle predictably at specific sites. Such variation or heterogeneity in the host plant and the predictable responses of parasites may allow the plant, over evolutionary and ecological time, to manipulate pests in such a way as to minimize their impact. Viewed in this light, patterns of host variation or even the absence of patterns may be key factors in the interactions between insects and their host plants and, as such, ecologists should perhaps be more concerned with them.

Acknowledgments. This paper was stimulated in part by discussions with Rodger Mitchell, whose elegant work with bean weevils helped refine my own views. I also benefited from discussions with Paul K. Dayton, Tom Gibson, Peter W. Price, C. N. Slobodchikoff and William V. Zucker. This work was supported by N.S.F. Grants DEB-7905848 and DEB8005602.

References

Baur, E.: Einführung in die experimentelle Vererbungslehre. Berlin, 1930.

Dermen, H.: Nature of plant sports. Am. Horti. Mag. 39, 123-174 (1960).

Dixon, A. F. G.: The role of aphids in wood formation. I. The effect of the sycamore aphid, *Drepanosiphum platanoides* (Schr.) (Aphididae), on the growth of sycamore, *Acer pseudoplatanus* (L.). J. Appl. Ecol. 8, 165-179 (1971a).

Dixon, A. F. G.: The role of aphids in wood formation. II. The effect of the lime aphid, *Eucellipterus tiliae* L. (Aphididae), on the growth of lime, *Tilia* x *vulgaris* Hayne. J. Appl. Ecol. 8, 393-399 (1971b).

Dixon, A. F. G.: Biology of Aphids. London: Camelot Press, 1973.

Fedde, G. F.: Impact of the balsam wooly aphid (Homoptera: Phylloxeridae) on cones and seed produced by infested Fraser fir. Can. Entomol. 105, 673-680 (1973).

Feeny, P.: Seasonal changes in oak leaf tannins and nutrients as a cause of spring feeding by winter moth caterpillars. Ecology 51, 565-581 (1970).

Feeny, P.: Plant apparency and chemical defense. In: Biochemical Interactions Between Plants and Insects. Rec. Adv. Phytochem., Vol. 10. Wallace, J. W., Mansell, R. L. (eds.). London: Plenum Press, 1976, pp. 1-40.

Fretwell, S. D.: Populations in a Seasonal Environment. Princeton, N.J.: Princeton Univ. Press, 1972.

Fretwell, S. D., Lucas, H. L., Jr.: On territorial behavior and other factors influencing habitat distribution in birds. I. Theoretical development. Acta Biotheor. 19, 16-36 (1970).

Gibb, J.: Predation by tits and squirrels on the eucosmid, *Ernarmonia conicolana* (Heyl.). J. Anim. Ecol. 27, 375-396 (1958).

Gibb, J.: Tits and their food supply in English pine woods: a problem in applied ornithology. In: Festschrift Vogelschutzvarte Hessen, Rheinland-Pfalz und Saarland, 1962, pp. 58-66.

Grant, V.: Genetics of Flowering Plants. New York: Columbia Univ. Press, 1975.

Green, T. R., Ryan, C. A.: Wound-induced proteinase inhibitor in plant leaves: A possible defense mechanism against insects. Science 175, 776-777 (1972).

Gustafson, G., Ryan, C. A,: Specificity of protein turnover in tomato leaves. J. Biol. Chem. 251, 7004-7010 (1976).

Hallé, F., Oldeman, R. A. A., Tomlinson, P. B.: Tropical Trees and Forests. New York: Springer-Verlag, 1978.

Harper, J. L.: Population Biology of Plants. London: Academic Press, 1977.

Hartmann, H. T., Kester, D. E.: Plant Propagation. Englewood Cliffs, N.J.: Prentice-Hall, 1975.

Haukioja, E., Niemelä, P.: Does birch defend itself actively against herbivores? Rep. Kevo Subarct. Res. Sta. 13, 44-47 (1976).

Haukioja, E., Niemelä, P.: Retarded growth of a geometrid larva after mechanical damage to leaves of its host tree. Ann. Zool. Fenn. 14, 48-52 (1977).

Haukioja, E., Niemelä, P.: Birch leaves as a resource for herbivores: Seasonal occurrence of increased resistance in foliage after mechanical damage of adjacent leaves. Oecologia 39, 151-159 (1979).

Hori, K.: Plant growth-promoting factor in the salivary gland of the bug, *Lygus disponsi.* J. Insect Physiol. 20, 1623-1627 (1974).

Keen, N. T., Simms, J. J., Erwin, D. C., Rice, E., Partridge, J. E.: 6α-Hydroxyphaseollin: an antifungal chemical induced in soybean hypocotyls by *Phytophthora megasperma* var. *sojae.* Phytopathology 61, 1084-1089 (1971).

Kozlowski, T. T.: Growth and Development of Trees, Vol. 1. New York: Academic Press, 1971.

Kozlowski, T. T., Ward, R. C.: Shoot elongation characteristics of forest trees. Forest Sci. 7, 357-368 (1961).

Kuć, J.: Phytoalexins. Ann. Rev. Phytopathol. 10: 207-232 (1972).

Llewellyn, M.: The effects of the lime aphid, *Eucallipterus tiliae* L. (Aphididae) on the growth of the lime, *Tilia* x *vulgaris* Hayne. I. Energy requirements of the aphid population. J. Appl. Ecol. 9, 261-282 (1972).

Llewellyn, M.: The effects of the lime aphid, *Eucallipterus tiliae* L. (Aphididae) on the growth of the lime, (*Tilia* x *vulgaris* Hayne). II. The primary production of saplings and mature trees, the energy drain imposed by the aphid populations and revised standard deviations of aphid populations energy budgets. J. Appl. Ecol. 12, 15-23 (1975).

Lurie, S., Paz, N., Struch, N., Bravado, B. A.: Effects of leaf age on photosynthesis and photorespiration. In: Photosynthesis and Plant Development. Marcelle, R., Clijsters, H., Von Poucke, M. (eds.). Boston: Dr. W. Junk Publ., 1979, 376 pp.

MacFarland, D., Ryan, C. A.: Proteinase inhibitor-inducing factor in plant leaves. Plant Physiol. 54, 706-708 (1974).

Matta, A.: Microbial penetration and immunization of uncongenial host plants. Ann. Rev. Phytopathol. 9, 387-410 (1971).

Miles, P. W.: Studies on the salivary physiology of plant-bugs: Experimental induction of galls. J. Insect Physiol. 14, 97-106 (1968).

Mittler, T. E.: Studies on the feeding and nutrition of *Tuberolachnus salignus* (Gmelin) (Homoptera, Aphididae). II. The nitrogen and sugar composition of ingested phloem sap and excreted honeydew. J. Exp. Biol. 35, 74-84 (1958).

Müller, K. O.: The nature of resistance of the potato plant to blight–*Phytophthora infestans*. J. Nat. Inst. Agr. Bot. 6, 346-360 (1953).

Neilson-Jones, W.: Plant Chimeras. London: Methuen, 1969.

Oinonen, E.: Sporal regeneration of bracken in Finland in the light of the dimensions and age of its clones. Acta Forest Fenn. 83, 3-96 (1967a).

Oinonen, E.: The correlation between the size of Finnish bracken (*Pteridium aquilinum* (L.) (Kuhn) clones and certain periods of site history. Acta Forest Fenn. 83, 1-51 (1967b).

O'Rourke, F. L.: Honeylocust as a shade and lawn tree. Am. Nurseryman 90, 24-29 (1949).

Rhoades, D. F., Cates, R. G.: Towards a general theory of plant antiherbivore chemistry. In: Biochemical Interactions between Plants and Insects. Rec. Adv. Phytochem., Vol. 10. Wallace, J. W., Mansell, R. L. (eds.). London: Plenum Press, 1976, pp. 168-213.

Richardson, S. D.: In: The Physiology of Forest Trees. Thimann, K. V. (ed.). New York: Ronald Press, 1958.

Shamel, A. D., Pomeroy, C. S.: Bud mutations in horticultural crops. J. Hered. 27, 486-494 (1936).

Shields, L. M.: Leaf zeromorphy as related to physiological and structural influences. Bot. Rev. 16, 399-447 (1950).

Spenser, P. W., Titus, J. S.: Apple leaf senescence: leaf disc compared to attached leaf. Plant Physiol. 51, 89-92 (1973).

Stewart, R. N., Dermen, H.: Ontogeny in monocotyledons as revealed by studies of the developmental anatomy of periclinal chloroplast chimeras. Am. J. Bot. 66, 47-58 (1979).

Vasek, F. C.: Creosote bush: long-lived clones in the Mojave Desert. Am. J. Bot. 67, 246-255 (1980).

Way, M. J.: Intraspecific mechanisms with special reference to aphid populations. In: Insect Abundance. Southwood, T. R. E. (ed.). Symp. Royal Entomol. Soc., London, No. 4. Oxford: Blackwell Scientific Press, 1968.

Way, M. J., Cammell, M. E.: Aggregation behavior in relation to food utilization by aphids. In: Animal Populations in Relation to their Food Resources. Watson, A. (ed.). Symp. Brit. Ecol. Soc., No. 10. Oxford: Blackwell Scientific Press, 1970.

White, T. C. R.: An index to measure weather-induced stress of trees associated with outbreaks of psyllids in Australia. Ecology 50, 905-909 (1969).

White, T. C. R.: The importance of a relative shortage of food in animal ecology. Oecologia 33, 71-86 (1978).

Whitham, T. G.: Habitat selection by *Pemphigus* aphids in response to resource limitation and competition. Ecology 59, 1164-1176 (1978).

Whitham, T. G.: Territorial behaviour of *Pemphigus* gall aphids. Nature (London) 279, 324-325 (1979).

Whitham, T. G.: The theory of habitat selection: examined and extended using *Pemphigus* aphids. Am. Nat. 115, 449-466 (1980).

Whitham, T. G.: Within-plant variation as a defense against parasites and herbivores. (1981) (in prep.).

Whitham, T. G., Slobodchikoff, C. N.: Evolution by individuals, plant-herbivore interactions, and mosaics of genetic variability: the adaptive significance of somatic mutations in plants. Oecologica (1981) (in press).

Woodward, R. G., Rawson, H. M.: Photosynthesis and transpiration in dicotyledonus plants. II. Expanding and senescing leaves of soybean. Austr. J. Plant Physiol. 3, 257-267 (1976).

Woodwell, G. M., Whittaker, R. H., Houghton, R. A.: Nutrient concentrations in plants in the Brookhaven oak-pine forest. Ecology 56, 318-332 (1975).

Zimmermann, M. H., Brown, C. L.: Trees: Structure and Function. New York: Springer-Verlag, 1971.

Zucker, W. V.: How aphids choose leaves: the role of phenolics in host selection by galling aphids. (1981) (in prep.).

Chapter 2

Responses of Black Pineleaf Scales to Host Plant Variability

GEORGE F. EDMUNDS, JR. and DONALD N. ALSTAD

1 Introduction

Plants use various means to deter or dissuade attack by phytophagous insects. Herbivore counteradaptation to these defenses is probably an important factor in insect host specificity and the coevolution of plant and herbivore phylogenies. This coevolutionary interaction is particularly interesting among long-lived forest trees and their associated insect fauna. One would think that the longevity of trees is a severe disadvantage in the coevolutionary interaction with short-lived insects; their relative success in spite of such herbivores invites further inquiry.

At least three mechanisms are likely to contribute to the success of forest trees against insects: First, counteradaptation to some plant defenses, notably the tannins and tannin-like compounds (Feeny 1976, Rhoades and Cates 1976), is very difficult. Many insects minimize exposure through the seasonal timing of their activity, feeding on young, low-tannin leaves (Feeny 1968, 1970) but, to the extent that successful counteradaptation is impossible, coevolution plays only a minor role in the interaction. Secondly, insects feeding on a predictable arboreal resource may be vulnerable to parasitism and predation, and interaction with predators and parasites is often an overwhelming determinant of herbivore population dynamics. Finally, trees are among "the world's most genetically variable organisms" (Antonovics 1980, see also Hamrick et al. 1979).

In the present paper we focus on this third phenomenon: we will describe coevolutionary patterns of tree variability and insect adaptation which have developed in the interaction of the ponderosa pine (*Pinus ponderosa* Lawson) and black pineleaf scale (*Nuculaspis californica* Coleman) insects.

2 Natural History of the Pine/Scale Interaction

Black pineleaf scale occurs on at least 11 species of *Pinus* and on Douglas fir (*Pseudotsuga*). Each outbreak of this scale is confined largely to one host species, with only rare individuals of other species attacked. Outbreaks are scattered and only a few persist for long periods. All outbreaks known to us (British Columbia, Washington, Oregon, California and Nevada) are caused when dust or drifting insecticide kills the Eulophid parasitoid (*Prospaltella* n. sp. near *aurantii* Howard) which normally suppresses scales to very low abundance. Association of black pineleaf scale outbreaks with "industrial fumes" or "smogs" (Furniss and Carolin 1977) appears to result similarly from the dust in such polluted air masses (Edmunds 1973). Outbreaks terminate when dust sources are controlled or when rapid onset of freezing weather catches the feeding scales before they become dormant. Spray practices influence outbreaks caused by insecticide drift; scale numbers generally decline when shifts are made to short residual insecticides, unless they drift in quantity from aerial applications.

In Spokane, Washington, black pineleaf scale has persisted at high densities from 1947 or earlier, and the current outbreak area is caused by local urban dust sources (automobile traffic on streets, alleys and parking lots). The study site area exists in a narrow elevational zone (625-637 m ground level) where significant winter-kill of the scale has not occurred since 1948. At lower elevations winter-kill is frequent and at higher elevations populations are reduced in episodes at approximately five-year intervals (Edmunds 1973).

Black pineleaf scale has one generation per year at our Spokane study site. It is reported to have two or three generations per year in southern California (Furniss and Carolin 1977), but we have been unable to confirm this observation on either ponderosa or Jeffrey pines. At Spokane, females have up to 40 eggs, and hatching extends over a 10-15-day period beginning about July 10th. Eggs hatch either in the oviduct, or immediately on leaving the female. We have never seen eggs under scale coverings, but we have observed first-instar larvae with the egg chorion attached. On ponderosa pine the new needle crop is within two weeks of complete elongation when the larvae begin to hatch. Larvae or crawlers settle on each of the annual needle crops, but most of them move to the youngest needles. A small percentage of the larvae is carried in the air to new trees, and we have caught some on sticky traps up to 0.75 km from the trees. Crawlers settle within two or three days and establish a thin white covering. They continue to feed until cold weather induces dormancy in late October or early November. There is some evidence that they feed during warm periods in March or early April. Continuous feeding resumes in mid-April and males are on the wing in mid-May. The sex ratio is strongly skewed, with females outnumbering males on average by about 10 to 1, but the ratio varies among trees.

All outbreaks of black pineleaf scale are characterized by extreme tree-to-tree variation in insect population density, and even in heavily infested stands pines can be found which remain completely free of scale. The branches of heavily infested trees may be intermingled with those of uninfested trees. Although first-instar scale larvae reach these trees, sometimes in fair numbers, they fail to survive, indicating that such trees are "resistant." Some form of "resistance" is also suggested by the fact that relative population density on a pine tends to remain for several years at about the same level in relation to neighboring trees. This "resistance," however, is less than immunity.

Repeatedly such individuals in stands numbered and monitored over a 20-year period (by Edmunds) have been successfully colonized. Subsequently, the infestation density grows slowly, requiring many generations before a heavy population inhabits the host. Every infested tree probably undergoes a similar ontogeny. Within stands, there is a strong association between infestation density and tree size (which at any site is correlated with age) (Fig. 2-1a).

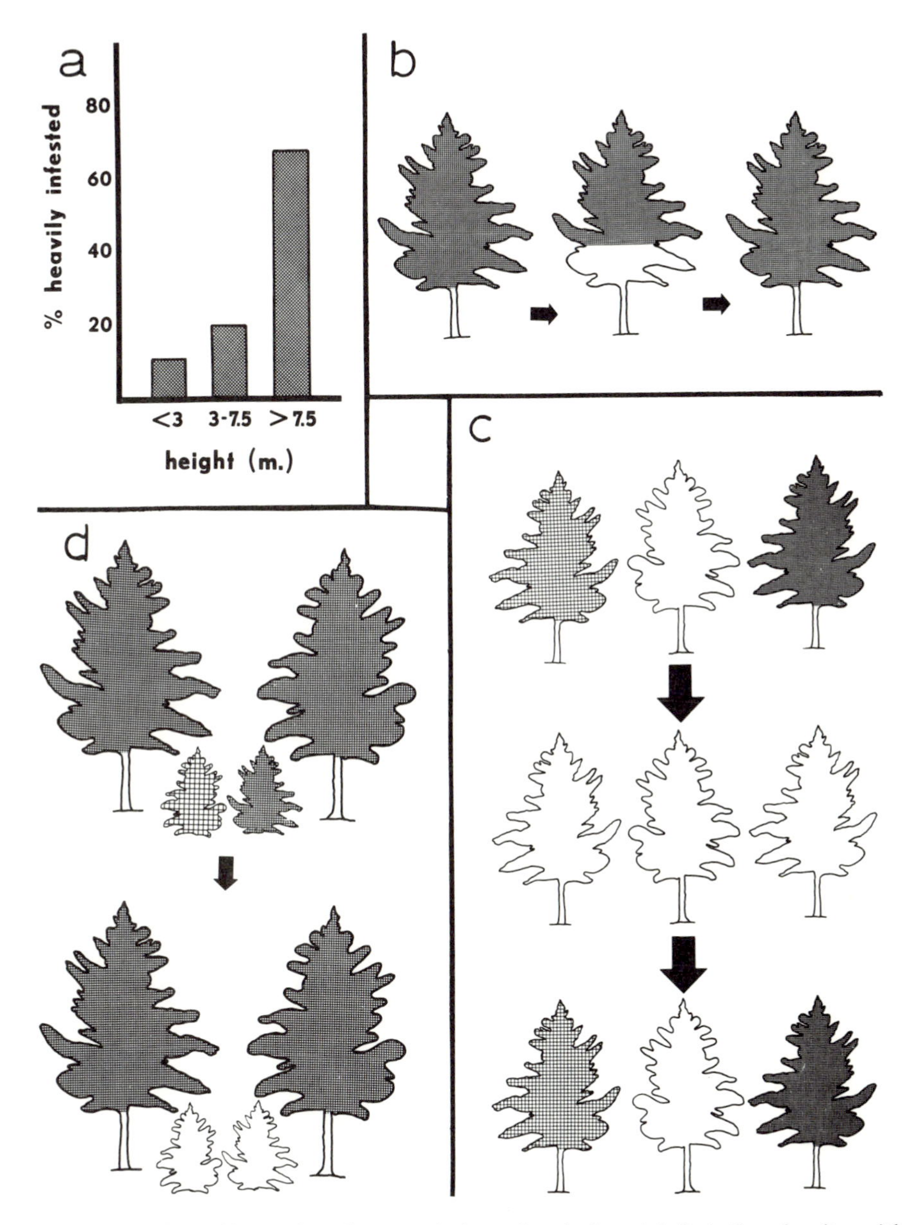

Figure 2-1. a: Data illustrating the association of scale insect infestation density with tree size and age. b, c, d: Diagrammatic representation of insecticide spray experiments further described in the text.

Simple insecticide spray experiments contribute insights into the pine/scale interaction. Malathion or Guthion treatment of infested plots can produce $>99\%$ scale mortality but, when the scale populations recover in 2-5 years, each tree carries the same relative population density it did previously, for example, formerly severely infested pines become severely infested (Fig. 2-1c), formerly lightly infested trees become lightly infested, and the uninfested trees remained uninfested. If instead of spraying the insecticide every twig is dipped in solution, 100% scale control can be achieved. When twigs on the lower limbs of a severely infested pine are dipped they are reinfested within 2 years (two generations) (Fig. 2-1b). In contrast, when 100% control was achieved on a group of small trees (1-2 m high) that were overgrown by trees densely infested with scale, 4 out of 5 trees remained scale-free for 10 years (after which they were destroyed during construction). The fifth tree exhibited the gradual reinfestation ontogeny described above (Fig. 2-1d). Differences in recolonization rate cannot be accounted for by the numbers of arriving larvae.

These observations and experiments suggest that ponderosa pine defenses are complex and variable, and that black pineleaf scale populations undergo selection and adaptation to the characteristics of an individual host (Edmunds 1973).

3 Transfer Experiments

We have explored the interaction of black pineleaf scale and pines with a series of field experiments which take advantage of the insects' sessile habit. Specifically, we hypothesized that: (1) individual trees vary in the defensive phenotypes which they present to scale insects; (2) selection over many generations produces scale insect populations which are increasingly adapted to the defensive character of their host tree (i.e., the insects track individual hosts); and (3) selection of scale insects for maximum fitness on one tree is maladaptive with respect to the prospect of establishing colonies of offspring on trees of different defensive phenotype (Edmunds and Alstad 1978).

3.1 Transfer Procedure

Each experiment involved a series of scale insect transfers from infested trees ("donors") to uninfested trees ("receptors") in another area. Donor trees were selected from within a 600-meter radius in scattered stands at an elevation of approximately 628 meters on the north city limits of Spokane, Washington; the receptor trees are located about 25 km NNW in Deer Park, Washington at an elevation of 634 meters. The receptor trees lie in the same air drainage and are subjected to similar climate. Because the receptor area is relatively dust-free, scale populations are very low and tree stands have not been exposed to their feeding in large numbers.

Transfers are made as the first-instar larvae (crawlers) appear on the donor trees; exact timing is essential and the experiments are limited to a few days each year. Infested twigs are removed from the donor trees and their new needles, on which scale crawlers would normally settle, are trimmed away. One-year-old needles, carrying mature females and their hatching progeny, are thinned, leaving 12-20 needles estimated to carry about 100 females. It was not possible to determine the number of females present or the number of eggs per female because scale-cover removal kills the insect. The older needles are removed, wounds are treated with an anti-desiccant, and the stems are inserted into water-absorbent florist foam in a small polyethylene bag to prevent the twigs from drying up before all the crawlers are produced. Each twig is then labeled to identify the donor, receptor, and replicate number (Fig. 2-2). The donor twigs are fastened in the trees with twist ties in such a position that the first-instar larvae can crawl to the new needle crop of the receptor twig.

A week later, after hatching and establishment of the young scales is complete, each transfer is reexamined, and the number of successful colonists quantified by estimating population sizes and assigned to one of six ranks. Identical censuses made at subsequent intervals allowed us to calculate a survival index (the ratio of final rank abundance to the initial rank abundance) for each transfer. For example, a transfer whose rank abundance fell from an initial rank 4 to rank 2 after 9 months (overwintered second instar) had a survival index of 0.5. This data manipulation compensates for differences in the number of crawlers transferred. It may also obscure some density-dependent phenomena. Our transfers produce insect densities far below those sustained on a heavily infested donor, and we believe that crowding mortality is unlikely; nevertheless, we are pursuing questions of density dependence with additional field work.

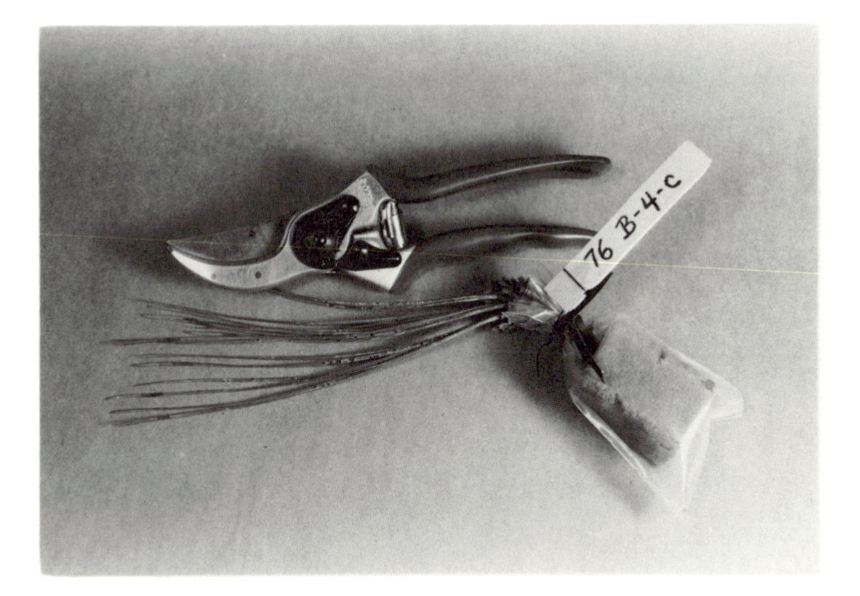

Figure 2-2. Infested twig from donor tree illustrating details of transfer sample preparation.

3.2 Between-Tree Analysis of Variance

Our first transfer experiment involved 10 donor trees and 10 receptors. Each possible combination was repeated in triplicate, resulting in a total of 300 transfers. Initial and 9-month (overwinter) abundance ranks were used to calculate a 10 X 10 X 3 array of survival indices suitable for the two-way analysis of variance (hereafter "ANOVA"). This statistical procedure allows us to test three aspects of the data set. The ANOVA main treatments examine differences within and between donors and receptors. Both results are highly significant, showing that the variance between scale demes originating from different donors is much greater than variance within scales from a single host (Table 2-1). Some scale demes are significantly better colonists than others, and this presumably is a result of their adaptation to different donor trees. Likewise, the receptor trees (columns) show statistically significant variation in "resistance." Scales do not survive as well on some trees as they do on others. The third test in the ANOVA, for interaction variance, is also significant; this is designed to detect exceptions to the main treatment patterns. Interaction would be suspected, for example, if insects from a relatively poor colonizing deme did especially well on a very resistant receptor tree. This result is also consistent with our hypothesis, which predicts that the success of scale transfers should correlate with the defensive similarity of the donor and receptor trees. It implies that resistance can only be quantified in the context of the insect genotypes against which a defense is mounted.

3.3 Within-Tree Controls

Concurrent with the age class transfer experiment, we also made replicated intratree control transfers to scale-free isolated branches that had been dipped in insecticide. Insects given the same experimental manipulation were allowed to establish in the

Table 2-1. Analysis of variance results. Accidental loss of some transfers resulted in unequal sample sizes in the treatment cells. Because partitioning order is important in nonorthogonal ANOVA designs of this type, two tables with reversed main treatment order are presented

Source	SS	DF	MS	F	P
Test of receptors preceding donors					
Within cells	173,091	172	1.006		
Receptors	48,976	9	5.442	5.407	0.001
Donors	19,980	9	2.220	2.206	0.024
Interaction	118,095	81	1.458	1.449	0.023
Test of donors preceding receptors					
Within cells	173,091	172	1.006		
Donors	20,888	9	2.321	2.306	0.018
Receptors	48,068	9	5.341	5.307	0.001
Interaction	118,095	81	1.449	1.449	0.023

Abbreviations: SS, sum of squares; DF, degrees of freedom; MS, mean square; F, F-ratio; P, probability.

trees from which they were taken. Six of the 20 transfers were subsequently destroyed by a construction project; data from the remaining 14 show 16 times as many larvae surviving to the second instar as do between-tree transfers.

3.4 Age Class Transfers

While our ANOVA results are consistent with the interpretation that selection causes differentiation of scale insect demes on individual pine trees, a competing counter-hypothesis remained. The association of tree age with scale population density could result simply from age-related weakening of tree defenses. To investigate this possibility, we conducted a transfer experiment on 10 donor trees and 10.receptors, with two replicates of each transfer. Receptors spanning a broad range of sizes were chosen, and 9-month survival indices were calculated. Means of the indices from each receptor showed no statistically significant correlation with receptor tree age. We therefore conclude that the strong relation between tree age and scale population density observed in outbreaks is not caused by a decline of resistance as the trees grow older.

3.5 Adaptation vs. Specificity Transfers

Our concern in the experiments described above was to determine whether insect demes were differentiated on individual trees. For this purpose, we chose heavily infested donor trees. Subsequently we have explored the phenomenon of scale host specificity by searching out correlates of the differentiation process. Specifically, we chose and ranked donors carrying a broad range of insect densities and asked the question "How does insect density, and presumably adaptation to the donor, correlate with average survival success in transfers to a variety of different receptors?" A detailed presentation of this recent experiment will be offered elsewhere (Alstad and Edmunds in prep.); suffice it to say here that the data show a significant inverse correlation ($N = 18$, $r_s = 0.56$, $P = 0.01$). Average survival of between-tree transfers to 10 different receptors declined as donor population density increased. Almost all of the mortality occurred early in development, before extensive feeding and needle chlorosis resulted (Fig. 2-3).

Our transfer experiments support the hypothesis that selection produces differentiated demes of black pineleaf scale on individual ponderosa pines. We have demonstrated the following: (1) Individual pines show variation that results in significantly differing survival success of scale insects artifically transferred to their branches. (2) Scale insects originating from separate donor trees show significant differences in the ability to survive on a variety of receptor trees. (3) Interaction variance in the transfer data is consistent with the hypothesis that survival success is correlated with the similarity of donor and receptor trees. (4) There is no evidence of age-related decline in tree defenses influencing insect survival. (5) Scale insect survival after intertree transfers is much lower than that of intratree control transfers. (6) Average between-tree colonizing ability declines as host adaptation and donor population density increase.

Figure 2-3. Receptor needles photographed 9 months after the transfer. One surviving third-instar scale is visible. In addition, scale covers of numerous insects which died in the first instar can be seen. At each site where a first-instar larva settled and began feeding, the needle is banded with a chlorotic lesion.

4 Discussion

We do not know which features of the pine chemistry are responsible for these patterns, but the diversity and variability of their "secondary" chemistries is well studied (Mirov 1968). They contain an array of terpenes, nonvolatile terpene acids, polyphenols and tannin-like substances, and the diversity and intertree variation of monoterpenes is so great as to give "almost every tree a unique individuality" (Hanover 1975, Sturgeon 1979, see also Smith 1964). This led us to speculate that terpene variation was the cause of the interaction observed in our experiments. In fact, scales feed in cortical cells well-removed from the resin ducts where terpenes are concentrated. The compounds that are involved must vary in much the same way as do the terpenes.

Whatever mechanisms are involved, intraspecific variation in ponderosa pines clearly influences scale insect fitness. Because of the difference in generation times, selection can and does produce scale populations which track individual host phenotypes. In this context, variability per se becomes an important feature of tree defenses; intraspecific defensive variation forces the coevolved herbivore into an evolutionary dilemma. When clear conflict exists between the ability to colonize new host individuals and adaptation to a single host, scale insects are forced to compromise.

Longevity is one of the principle life history parameters which correlates directly with electrophoretically detectable variability in forest trees (Hamrick et al. 1979).

While many factors, including large ranges, population size, and overlapping generations must contribute to this pattern, we feel that the frequency-dependent interaction with short-lived herbivores may be important in balancing polymorphic tree populations. The same interaction may contribute to the legendary instability of forest insect densities. If successful intertree colonizations require a minimum level of preadaptation, any circumstance which changes insect abundance will mean that more or fewer insect genotypes are tried on more or fewer trees.

Because of their sessile habit, black pineleaf scale insects have been convenient subjects in our study of intraspecific pine variation. Other scales may reflect similar interactions with their hosts, e.g., *Toumeyella numismaticum* (P. and McD.) (Coccidae) (Orr 1931) and *Matsucoccus resinosae* (Bean and Goodman) (Margarodidae) (McClure 1977). We do not believe, however, that the sedentary habit is a fundamental independent variable in the interaction. While it is possible that local microdifferentiation would not occur among more mobile insects, the relevant question is not whether they can move, but whether they do move, and how much gene flow occurs between insect population units on different trees. Many forest insects display adaptations which suggest commitment to individual host trees. Why are there so many examples of wingless females among forest moths of the families Psychidae, Geometridae and Lymantriidae? Why do many winged female Lepidoptera oviposit on their cocoon? Why are others too heavy to fly until they deposit a fraction of their eggs (Baker 1972)? Why do spruce budworm outbreaks characteristically occur in mature and overmature tree stands (Morris 1963)?

Because we perceive variability per se as an important factor in the interaction of trees, especially conifers, and herbivorous insects, we regard the silvicultural practices of inbreeding, selection, and vegetative propagation of fast growth genotypes, which reduce variation in the tree stand, as unwise. Ultimately we may learn to manipulate the planting pattern of a variety of genotypes to maximize dissimilarity of neighbors, but a highly effective artificial system will depend on a much better understanding of tree variation and its perception by insect herbivores.

References

Alstad, D. N., Edmunds, G. F., Jr.: Adaptation and host specificity of black pineleaf scale insects (1981). (in prep.).

Antonovics, J.: The study of plant populations. Science 208: 587-589 (1980).

Baker, W. L.: Eastern forest insects. U.S. Department of Agriculture Forest Service Misc. Publ. No. 1175 (1972).

Edmunds, G. F., Jr.: The ecology of black pineleaf scale. Environ. Entomol. 2, 765-777 (1973).

Edmunds, G. F., Jr., Alstad, D. N.: Coevolution in insect herbivores and conifers. Science 199, 941-945 (1978).

Feeny, P. P.: Effect of oak leaf tannins on larval growth of the winter moth *Operophtera brunata*. J. Insect Physiol. 14, 805-817 (1968).

Feeny, P. P.: Seasonal changes in oak leaf tannins and nutrients as a cause of spring feeding by winter moth caterpillars. Ecology 51, 565-581 (1970).

Feeny, P. P.: Plant apparency and chemical defense. Rec. Adv. Phytochem. 10, 1-40 (1976).

Furniss, R. L., Carolin, V. M.: Western forest insects. U.S. Department of Agriculture Forest Service Misc. Publ. 1339 (1977).

Hamrick, J. L., Linhart, Y. B., Mitton, J. B.: Relationships between life history characteristics and electrophoretically detectable genetic variation. Ann. Rev. Ecol. Syst. 10, 173-200 (1979).

Hanover, J. W.: Physiology of tree resistance to insects. Ann. Rev. Entomol. 20, 75-95 (1975).

McClure, M. S.: Populations dynamics of the red pine scale *Matsucoccus resinosae:* the influence of resinosis. Environ. Entomol. 6, 789-795 (1977).

Mirov, N. T.: The Genus *Pinus.* New York: Ronald, 1968.

Morris, R. F.: The dynamics of epidemic spruce budworm populations. Memoirs of the Entomological Society of Canada No. 31 (1963).

Orr, L. W.: Studies on natural vs. artificial control of the pine tortoise scale. Univ. Minn. Agric. Exp. Stat., Tech. Bull. 79, 1-19 (1931).

Rhoades, D. F., Cates, R. G.: Toward a general theory of plant antiherbivore chemistry. Rec. Adv. Phytochem. 10, 168-213 (1976).

Smith, R. H.: Variation in the monoterpenes of *Pinus ponderosa* Laws. Science 143, 1337-1338 (1964).

Sturgeon, K. B.: Monoterpene variation in ponderosa pine xylem resin related to western pine beetle predation. Evolution 33, 803-814 (1979).

Chapter 3

The Role of Host Plants in the Speciation of Treehoppers: An example from the *Enchenopa binotata* Complex

Thomas K. Wood and Sheldon I. Guttman

1 Introduction

Recent studies on the origin of insect races and speciation in certain phytophagous insect complexes have forced systematists to revise their ideas concerning modes of animal speciation. Rapid establishment of new races by insects on introduced plants has led some biologists to suggest that races and species may arise sympatrically (Bush 1969, 1975). Since much of the evidence is indirect, however, other workers still regard geographical isolation as a prerequisite for speciation in most groups of sexually reproducing animals (Mayr 1970).

Phytophagous and parasitic insects are particularly suited to sympatric divergence because of specialized feeding habits (Bush 1975). Maynard Smith (1966) showed that in theory disruptive selection operating on genotypes differing in habitat or host selection could lead to sympatric speciation. Allochronic life histories (Alexander and Bigelow 1960, Alexander 1968) or adoption of new host plants (Bush 1969, Huettel and Bush 1972, Knerer and Atwood 1973) have been implicated in the formation of insect races and reproductive isolation. Edmunds and Alstad (1978; Chapter 2) have demonstrated that sessile scale insects adapt to the defense system of individual pine trees. When dispersal abilities are limited and extinction rates high, newly colonized hosts may act as islands, with the insect population on one host plant partially isolated from demes on other adjacent trees (Simberloff 1976). Recently, Tauber and Tauber (1977a, 1977b) and Tauber et al. (1977) suggested a basis for sympatric divergence in a non-host-specific insect through selection for genes controlling diapause. These studies argue that sympatric speciation may have occurred, but, as Bush (1969) suggests, there is still a general paucity of detailed studies.

The Membracid, *Enchenopa binotata* (Fig. 3-1), has been historically considered as a single species. It occurs on a number of different host plants that occur throughout a broad geographical area from eastern North America to as far south as Panama (Metcalf and Wade 1965). Although very little is known about its tropical biology, host plant records indicate that it may be polyphagous in Central America. In Costa Rica, the species has been recorded on 13 host plants belonging to 12 genera in 6 families

(Ballou 1936). The distribution of *E. binotata* in eastern North America closely coincides with that of its 7 primary host plants: *Ptelea trifoliata* L. (hoptree); *Celastrus scandens* L. (bittersweet); *Robinia pseudoacacia* L. (black locust); *Cercis canadensis* L. (redbud); *Juglans nigra* L. (black walnut); *J. cinerea* L. (butternut); and *Viburnum* sp. Although these plants are sympatric over a wide geographic area, they seldom all occur in the same habitat, except where planted together as ornamentals.

Here we review studies showing that *E. binotata* in North America is a complex of reproductively isolated taxa which have diverged along host lines. Our objective in these studies has been to determine how host plants shape the life histories and promote speciation in the *E. binotata* complex. Wood (1980) postulated: (1) that the *E. binotata* progenitor was tropical, multivoltine and polyphagous, (2) that successful colonization of deciduous north temperate hosts resulted in coordination of egg-hatch with host phenology, (3) that the coordination of egg-hatch to host phenology on diverse hosts resulted in allochronic life histories and assortative mating, (4) that assortative mating was facilitated further by low vagility, female monogamy, habitat heterogeneity and ant mutualism, and (5) that the interaction of these factors promoted genetic polymorphism and reproductive isolation.

2 Study Area and Methods

All the studies reported here (electrophoretic studies excepted) were done with *Enchenopa* from Clinton County, Ohio. Our main field study area consisted of a 300 X 180 m site in the city of Wilmington, where treehopper populations on all hosts (except *C. scandens*) have coexisted for at least 10 years. In this area there were 156 individual host trees with branches from several tree species often interdigitating. Tree

Figure 3-1. Female of *Enchenopa binotata* on *Cercis canadensis* covering her egg mass with froth. Eggs are inserted into twigs.

heights varied from under 7 m to 24 m depending on the species. Field studies on associated ants and the phenology and vagility of treehoppers were done at this site. Samples for electrophoresis were collected from this area, nearby areas in Clinton County, and also from Wooster and Oxford, Ohio.

Differences among the life histories, host plant selection and mate selection of *Enchenopa* on each host species were studied in a small area (3.6 X 4.8 m) less than 1 km from the main study area. Each host plant species was caged and stocked with field-collected female treehoppers from original host plants. There they were allowed to oviposit. The phenology of egg-hatch on each host was determined the following spring by daily examination. The day eggs first hatched (day 1) was used as a reference for a continuous time scale to compare life history events of *Enchenopa* on the various host plants. Treehopper maturation, adult survival, and the phenology of dispersal, precopulation and copulation were measured by making daily observations over a 102-day period with each day divided into 14 observational periods beginning at 7:30 a.m. and ending at 8:30 p.m. Individual males and females were identified by marking with enamel paint. Analysis of Variance and Duncan's Multiple Range Test were used to compare means.

3 Life History Differences of *Enchenopa* on Host Plants

3.1 Treehopper Coloration and Feeding Site

Coloration of *Enchenopa* adults and nymphs varies with host species. Adult females and males on *R. pseudoacacia* and *Viburnum* are light to dark brown, while on other hosts they are black. Color variation as well as differences in feeding sites among *Enchenopa* on different hosts are most striking in third to fifth instars. Nymphs are black and white on the two *Juglans* species and contrast strikingly with the leaf petiole on which they feed. On *R. pseudoacacia,* nymphs also feed on the petiole but are green. On *Viburnum, C. canadensis* and *C. scandens* nymphs are brown to gray, and blend with the twigs of their hosts. Nymphs on *P. trifoliata* feed on twigs, and are a grayish-brown with a yellow stripe on the dorsum (Wood 1980).

3.2 Oviposition Characteristics and Behavior

Females insert their egg masses into the bark of the various host plants. Egg masses contain a mean of 14.5 eggs on *R. pseudoacacia* and 7.0-8.6 eggs on the other hosts (Wood 1980). Following oviposition, egg froth, a white secretion from the ovipositor, is deposited over the mass of eggs (Fig. 3-1). The major component of egg froth is an ether-extractable lipid that comprises from 77.7 to 87.1% of the froth mass. The remainder of the egg froth mass is protein (9.0-21.8%) and carbohydrate (1.5-3.6%). The proportion of these froth components varies and is dependent on the host plant (Wood 1980). In a series of experiments we found that the ether-extractable lipid in the egg froth contained an ovipositional attractant. Once an egg mass was deposited on a branch and covered with egg froth the ovipositional attractant drew other females to the same branch to deposit eggs. As the number of egg masses increased, more and more females were attracted to the branch. Consequently, egg masses are highly clumped and only a small proportion of the branches on an individual tree will have egg masses, but the ones that do can possess over 100 masses (Wood and Seilkop, unpubl.).

Egg froth deposited by females on different hosts has a distinctive color, shape and amount. When on the two *Juglans* species, the white egg froth shrinks and discolors shortly after deposition, becoming nearly indistinguishable from the host plant. Froth on the remaining host plants retains its color and shape.

Females on *J. nigra* and *R. pseudoacacia* deposit their eggs in new branches from the current growing season. Females on *P. trifoliata* and *Viburnum* deposit eggs in second-year growth, while those on *C. scandens* oviposit in 1-3-year-old twigs. The most dramatic difference occurs on *C. canadensis,* where females deposit eggs in 2-4-year-old growth. Branch diameter appears to be a critical factor influencing the site of oviposition (distance from terminal bud) as well as egg mass density. For instance, branches chosen for oviposition on *C. canadensis* have significantly greater diameters than those chosen on other hosts. Although egg masses can be found on first-year twigs with diameters similar to those of other host species, small branches suffer high overwinter die back in southern Ohio. The young twig mortality associated with *C. canadensis* has proably selected for oviposition on larger twigs.

Observations of treehoppers in the field and in cages showed that there are seasonal differences in oviposition patterns among females on different hosts (Fig. 3-2). Also, the time of day females deposit eggs varies with host. For instance, females on *P. trifoliata, C. canadensis, V. opulus* and *C. scandens* deposit the majority of their eggs in the morning, while females on *J. nigra* deposit eggs in the afternoon, and those on *R. pseudoacacia* deposit eggs in the evening (Wood 1980).

3.3 Egg-Hatch and Adult Maturation

The hatching phenology of eggs differed among hosts (Fig. 3-2). Eggs in *P. trifoliata, V. opulus* and *C. scandens* always began to hatch before those on *J. nigra, C. canadensis* and *R. pseudoacacia.* In some years eggs on *R. pseudoacacia* hatched after egg-hatch was nearly completed on the other host species. Also, the sequencing of egg-hatch on hosts remained consistent over 4 years of observation (Wood and Guttman 1981). Except for *C. canadensis,* eggs hatched on host plants when they were flowering. On *C. canadensis,* eggs hatched after flowering when leaves were fully formed.

In addition to disparate oviposition and hatching phenology, nymphs raised on different hosts molted to adults at different times (Fig. 3-2). Generally, nymphs molted to adults 30 days after egg-hatch and the sequence of adult maturation on host plants paralleled that of egg-hatch. Thus, differences in adult maturation time were largely attributable to staggered egg-hatch rather than inequitable nymphal development rate. Nymphs on *P. trifoliata, V. opulus* and *C. scandens* molted to adults first, followed by those on *J. nigra* and *C. canadensis,* and finally by those on *R. pseudoacacia.* Initial sex ratios following molting were 1:1 on all hosts. Thereafter, male survival fell consistently before mating and throughout the mating period. Following mating, only 18% of the initial cohort of males was alive on all hosts. Also, there were temporal differences

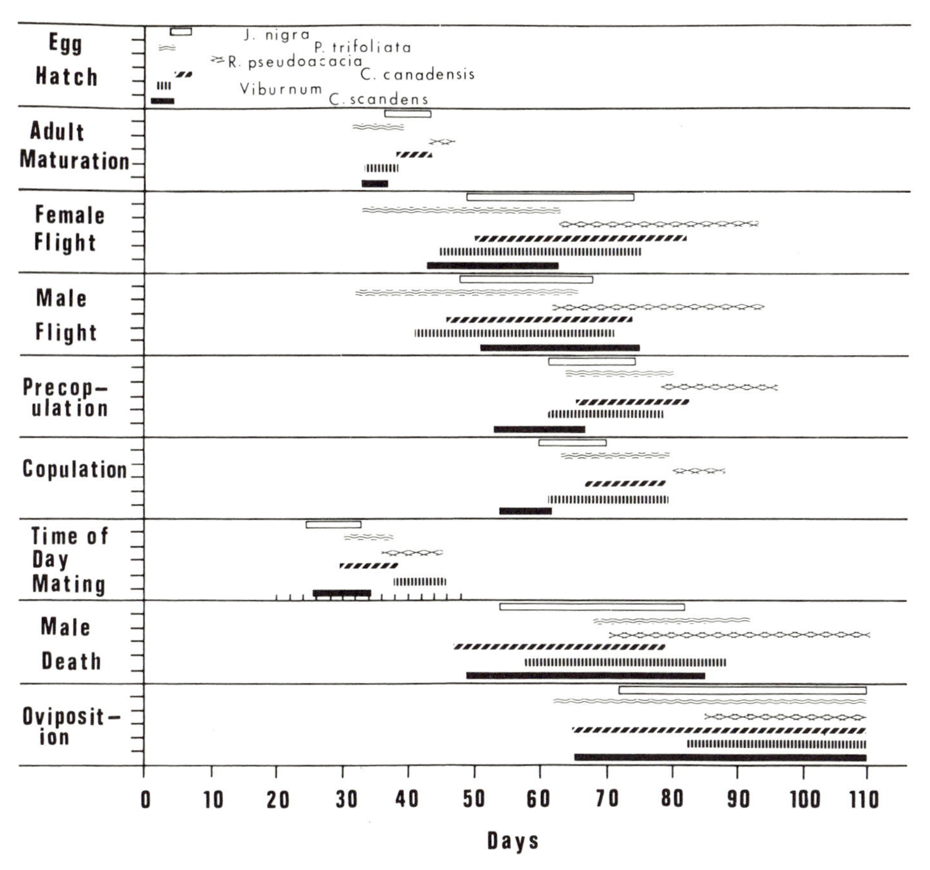

Figure 3-2. Phenological differences in egg-hatch, adult maturation, female and male dispersal (flight), precopulation, copulation, male mortality and oviposition among the host races of *Enchenopa*. Time scale is referenced to date of first egg-hatch. The time of day mating occurs is on a different scale. The day is divided into one-hourly intervals beginning at 7:30 a.m. and ending at 9:30 p.m.

in male treehopper mortality on the various host plants (Fig. 3-2). On all hosts the number of females decreased before and during mating, but mortality was not as great as that for males. By the end of the mating period, 76% of the female cohort was alive. Following mating, females lived from 30 to 60 days (Wood and Guttman 1981).
number of females decreased before and during mating, but mortality was not as great as that for males. By the end of the mating period, 76% of the female cohort was alive. Following mating, females lived from 30 to 60 days (Wood and Guttman 1981).

3.4 Adult Dispersal

We used the number of males and females gathered on the sides of the screen cages as an index of dispersal from host plants. Based on this index, adults dispersed from hosts at different times (Fig. 3-2), increasing the probability of meeting mates having developed on the same host. For example, females on all hosts but *J. nigra* and *V. opulus* differed in their times of dispersal. Similarly, males on *R. pseudoacacia* and *P. trifoliata* differed from each other and all other hosts in their dispersal behavior. Males on other hosts were not as clearly separated. Furthermore, treehoppers on different host plants displace one another by dispersing at different times of the day or at different temperatures (see Wood and Guttman 1981).

The majority of all *Enchenopa* were observed on the screens of cages before mating commenced. Three days after mating began, 63.2% of females and 52.2% of the males were recovered on the screens. Thus, a substantial amount of dispersal occurs prior to mating. For example, mating first occurred on *P. trifoliata* on day 55, and at that time, 72.2% of males and 84.5% of females had dispersed to the screen. Following mating and during oviposition, very few males or females dispersed (Wood and Guttman 1981). This experiment fails to differentiate between intra- and interhost dispersal but, as will be pointed out in the section that follows, the majority of dispersal is probably among conspecific hosts.

3.5 Vagility of Treehoppers

We hypothesize that low vagility acts in concert with low levels of intrahost dispersal to minimize contact among populations of *Enchenopa* on different hosts. A mark-recapture study was designed to estimate vagility. Adult *Enchenopa* were collected when the majority of nymphs on a given host species had molted to adults and before much flight activity had taken place. These were color-coded with enamel paint and reestablished on their original host. Counts of marked treehoppers were made daily on the release host and nearby plants for the first 4-5 days. Thereafter, counts were made every 2-3 days through mid-September (Wood and Guttman 1981).

The percentage of males and females recaptured at their site of release decreased rapidly during the first 17 days. During this period 7-63% of 217 released males were recovered on the release hosts and only 1-2% on nonhost trees. This time interval corresponded to the period of high mortality incurred during flight and before mating in the cage experiment. In the field, only a few marked males were observed on *Enchenopa* hosts other than their developmental ones. These males were observed 18.2 m from their release site and the remainder were observed within 4.5 m. By the time mating commenced in the cages, 6% of the released males were recaptured on their original host. After mating was completed on conspecific hosts in cages, very few marked males were observed in the field (Wood and Guttman 1981).

The decline in the number of marked females at release sites in the field also corresponded to the pattern of early mortality and flight activity observed in cages. The number of marked females dropped to zero 31-56 days after release on *J. nigra, C. scandens, C. canadensis* and *R. pseudoacacia.* This was not surprising considering the number of individual conspecific hosts and female mortality. However, observations of marked females on *P. trifoliata* and *V. prunifolium* demonstrated the tendency of females to remain at or near the release site. Between 10 and 13% of 88 marked females were observed on these hosts 43-64 days after release, respectively. If these data are adjusted using mortality figures from cage studies, they suggest that during this period 35.0-41.4% of the females reared on a given tree remain to deposit eggs. When male recaptures are adjusted for mortality the data suggest that 5-20% of the males on a given tree remain to the end of the mating period. These adjusted values are conservative, since mortality in the field is probably much greater than in the cages. Throughout this study marked *Enchenopa* were observed forming precopulatory and mating pairs on the release tree, and females were also observed depositing eggs. Our data suggests that populations of *Enchenopa* exhibit low vagility, and that the dispersal which does occur is largely intrahost rather than interhost (also see Wood and Guttman 1981).

3.6 Mating Activity

We recognize precopulatory behavior and copulation as two separate activities. Precopulatory behavior consisted of males sitting on females and lasted from a few seconds up to two hours. Precopulatory behavior may result in copulation, but males also abandon females and move to others. Precopulatory pair formation provided a measure of when males on different hosts became sexually active.

Precopulatory pairs began to form first during the season on *C. scandens,* followed in succession by those on *P. trifoliata, J. nigra, V. opulus, C. canadensis* and *R. pseudoacacia* (Fig. 3-2). The mean day precopulatory pairs formed on all host plant species differed significantly except for those on *V. opulus* and *J. nigra.* Also, the time of day that pairs formed on *C. scandens* and *J. nigra* was the same as it was for *Enchenopa* on *P. trifoliata, C. canadensis* and *J. nigra.* However, *Enchenopa* on *V. opulus* and *R. pseudoacacia* differed from each other and from those on all other hosts in the time of day they formed precopulatory pairs (Wood and Guttman 1981).

Treehoppers on different hosts displaced one another by copulating at different times in the season (Fig. 3-2). The mean day *Enchenopa* copulations took place on *C. scandens, J. nigra* and *R. pseudoacacia* differed significantly from each other and all other hosts. Mean day of mating for treehoppers on *V. opulus, P. trifoliata* and *C. canadensis* did not differ from each other. The time of day copulations occurred further divided *Enchenopa* on some host plants (Fig. 3-2). Treehoppers on *V. opulus* and *R. pseudoacacia* mated in the evening and differed from those on other hosts. Thus, *Enchenopa* on all hosts except *P. trifoliata* and *C. canadensis* mated at either different times of the season or day (see Wood and Guttman 1981).

3.7 Competition for Mates

Females outnumbered males on all hosts at the end of the mating period. However, females generally mated only once, while males secured up to four copulations. Thus, as the number of receptive females decreases, competition for mates should increase. A consequence of this might be the dispersal of males from their host in search of females. However, we did not notice an increase in male flight activity as the number of virgins decreased. The lack of male dispersal may be explained by two facts: First, male mortality was high at the end of the mating period, reducing the pool of competing males. Second, males failed to differentiate between virgins and previously mated females. Apparently, male fitness is increased by remaining on hosts where females are already present. Also, temporal differences in male mortality (see Fig. 3-2) may reduce the probability for interhost matings, but these are probably the result of differential selective pressures associated with hosts that place constraints on male development and/or longevity.

4 Host Plant and Mate Selection

4.1 Host Plant Selection

In addition to the disparate life histories that *Enchenopa* obtains on its host plants, features of host plant and mate selection behavior may further promote reproductive isolation among host plant populations. Field observations suggest that females select and oviposit in the host species on which they develop. To test this hypothesis, *Enchenopa* were confined in large cages containing all 7 host plant species. *Enchenopa* from one host were placed on that host in the center of the cage with free access to the other 6 surrounding host plants. *Enchenopa* from all 7 host plants were tested in a similar manner. Males and females dispersed throughout the cages before mating, and considerable interhost movement was observed before oviposition. When females deposited egg masses, they selected their native host. There were very few ovipositional "mistakes" and most of these occurred between the two *Juglans* species. Despite small cages and exaggerated sympatry, it appears that females were able to select and oviposit on the host plant on which they developed (see Wood 1980).

4.2 Mate Selection

Using an experimental design similar to that above, we tested whether *Enchenopa,* when given free access to mates from all host plants, selected mates according to their particular host plant origin. Six of the 7 host species were covered with a single cage. Adult *Enchenopa* from each host plant were field-collected prior to mating and color-coded with enamel paint to indicate host origin. Hourly observations were made from 8:00 a.m. to 8:30 p.m. throughout the mating period. As in the previous experiment, males and females dispersed throughout the cage with considerable interhost movement before mating (Wood 1980, Wood and Guttman 1981).

Before mating occurred in *Enchenopa* there was a period of precopulatory pair formation during which males and females sat next to each other up to two hours. Precopulatory pair formation among males and females from different host plants was common (Wood 1980). In one year, 166 of 288 precopulatory pairs involved partners from the same host plant (Wood and Guttman 1981). These data suggest that males cannot distinguish receptive females on the basis of their host origin. However, of the 128 observed copulations, only 7 involved males and females of mixed plant origin. Several mixed copulations were of shorter duration than normal (< 142 min), suggesting that sperm may not have been transferred (Wood 1980). We conclude that females select males of the same host plant origin with which to mate.

5 Genetic Differentiation among Host Populations of Treehoppers

Determination of intra- and interhost differentiation was done by examining 15 presumptive loci with horizontal starch gel electrophoresis. Of these 15 loci, 7 were polymorphic and used for comparisons. Allele frequencies among *Enchenopa* indicate that all host plants of a given species support populations with identical predominant alleles at a given locus. However, the results of chi-square homogeneity tests demonstrate that significant differences in allele frequencies exist within most *Enchenopa* species. This genetic differentiation occurs at the microgeographic (individual tree) and macrogeographic levels as well as between developmental stages. The degree of differentiation varies depending upon the host population of *Enchenopa* being considered. For example, significant differences in allele frequency existed for only one locus in insects from individual *C. canadensis* in the main study area, while significant differentiation was present at 4 of the 6 variable loci tested in nymphs from *C. scandens* (Guttman et al. 1981).

The magnitude of genetic differentiation among *Enchenopa* host populations (Table 3-1) is much greater than within a host population (Guttman et al. 1981). Treehoppers on *J. nigra* are readily distinguished from the other host populations by the near fixation of a glutamate oxaloacetate transaminase allele. This allele is absent from most treehopper populations on other host plants but, if found, is present in low frequency (0.17 or less). Similarly, those *Enchenopa* populations on *P. trifoliata* are fixed, or nearly so, for a peptidase allele. Others which have this allele are populations on *J. nigra* and *C. canadensis,* where its frequency is 0.16 or less. A catalase allele is fixed in *Enchenopa* on *P. trifoliata,* while it occurs in low frequency in *J. nigra* and *C. canadensis* populations. Treehoppers on *R. pseudoacacia* possess a particular phosphoglucomutase allele in frequencies ranging from 0.62 to 0.75. This allele is absent in most other populations, but when it does occur its frequency does not exceed 0.11 (Guttman et al. 1981).

The statistic $\bar{D}$, as defined by Nei (1972) and modified by Nei (pers. comm.), was used to estimate the genetic distance within and among host populations of treehoppers. If it is assumed that the substitution of electromorphs occurs as a Poisson process, then $\bar{D}$ is the average number of substitutions per locus in the separate evolution of the two populations being compared. A total of 41 populations were analyzed. The mean genetic distance among the combined *Enchenopa* host populations is low ($\bar{D} = 0.11 \pm 0.06$) but is about three times the average distance of *Enchenopa* populations from only *J. nigra* ($\bar{D} = 0.04$), the most variable host population. The $\bar{D}$ between *Enchenopa* from *J. nigra* and the remaining host entities is 0.17 ± 0.05. The value for those from *P. trifoliata* is 0.13 ± 0.04, and the value for those on *R. pseudoacacia* is 0.13 ± 0.07. Treehoppers on *C. canadensis* show smaller distances than those from the previous three hosts. They maintain a $\bar{D}$ of 0.10 ± 0.04 relative to the other entities, and the smallest $\bar{D}$ between *C. canadensis* treehoppers and any other host population is 0.05, suggesting a large degree of genetic distinctness. Biochemically, *Enchenopa* on *Viburnum* and *C. scandens* ($\bar{D} = 0.01$) are very similar, suggesting that differentiation has been minimal (Guttman et al. 1981).

A dendrogram summarizing genetic distances and the history of divergence in the *E. binotata* complex was obtained following cluster analysis of $\bar{D}$ values by the un-

Table 3-1. Mean allele frequencies at 6 gene loci coding for enzymes in fifth-instar nymphs of *Enchenopa* on 6 host plants. All treehoppers were collected within a 10-km area in Ohio

Locus	Allele	*Juglans* (5) N = 150	*Ptelea* (4) N = 158	*Cercis* (4) N = 130	*Viburnum* (4) N = 140	*Robinia* (1) N = 26	*Celastrus* (5) N = 158
Est-1	*a*				0.01		
	b	0.08	0.05	0.05	0.19		0.15
	c	0.67	0.58	0.07	0.29		0.41
	d	0.25	0.18	0.80	0.39	1.0	0.29
	e		0.19	0.08	0.10		0.10
	f				0.02		0.05
Pep-2	*a*	0.15	0.99				
	b	0.53	0.01	0.34			0.34
	c	0.30		0.64	0.98	1.0	0.66
	d	0.01		0.02	0.02		
Pgi-2	*a*	0.20	0.14		0.01		
	b	0.77	0.86	0.98	0.93	0.92	0.98
	c	0.01		0.02	0.06	0.08	0.01
	d	0.01					0.01
Pgm-2	*a*	0.01	0.14	0.02	0.02		0.01
	b				0.01		
	c	0.95	0.86	0.94	0.96	0.35	0.92
	d	0.03		0.02		0.04	0.03
	e	0.01		0.02		0.62	0.04
Cat-1	*a*	0.45		0.02	0.03	0.04	0.03
	b			0.34			
	c	0.46		0.62	0.22		0.25
	d	0.09	1.0	0.02	0.73	0.96	0.70
	e				0.02		0.01
Got-1	*a*						0.02
	b	0.03	0.08	0.08	0.03		0.19
	c		0.83	0.92	0.97	1.0	0.78
	d	0.97	0.09				0.01

In parentheses are the numbers of trees from which samples were taken.
N = number of genomes sampled.
Code for alleles: Est, esterase; Pep, peptidase; Pgi, phosphoglucose isomerase; Pgm, phosphoglucomutase; Cat, catalase; Got, glutamate oxaloacetate transaminase.

weighted pair group method (Tateno, pers. comm.) (Fig. 3-3). The use of such a phenogram in making phylogenic inferences relies upon the assumption that there is a constant rate of gene substitution per unit length of time in all evolutionary branches. The dendrogram indicates that *Enchenopa* on *J. nigra* diverged first, followed by those on

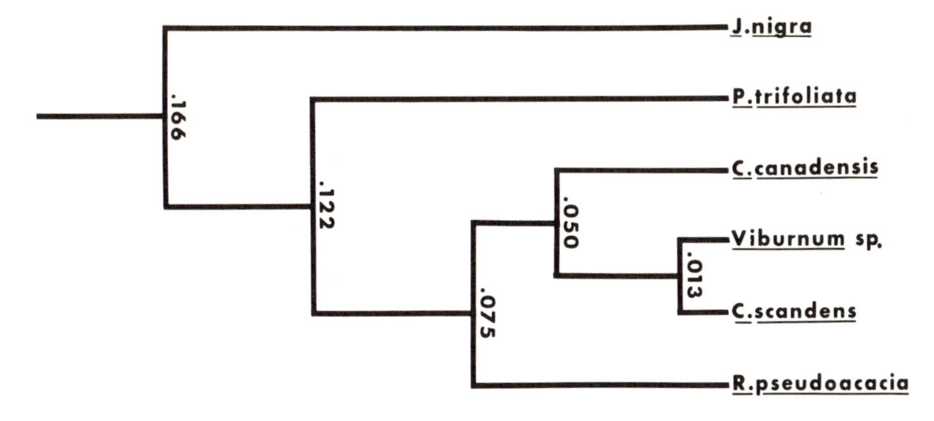

Figure 3-3. Dendrogram showing the relationships among the 6 host races of *Enchenopa* based on genetic distances.

P. trifoliata, R. pseudoacacia and *C. canadensis.* Treehoppers in the *Viburnum-C. scandens* group diverged most recently (Guttman et al. 1981). Finally, we feel that the genetic distances observed suggest that gene flow has been minimal among *Enchenopa* on some hosts and that reproductive isolation has occurred.

Finally, we will speculate on the time of divergence of these *Enchenopa* entities. Nei (1972) states that, if most of the biochemical changes detected are neutral or relatively neutral, then the rates of protein evolution would be stochastically constant, and protein differences would reflect not only phylogeny but could serve as an evolutionary clock indicating actual times the various cladogenetic events occurred. One need not be a strict "neutralist" to make use of Nei's model. Over all loci under selection, some will be evolving more conservatively, and others more rapidly, than the overall mean. If we examine a diversity of loci with differing evolutionary rates (Sarich 1977), it is to be hoped that we average the discrepancy in rates among individual loci. Genetic distance has been used as an estimate of divergence time in a number of studies, and times obtained appear to correlate well with estimates obtained from other evidence, such as geological history (Gorman et al. 1971, 1975, Nevo et al. 1974, Yang et al. 1974, Kim et al. 1976). However, several different methods have been used to calibrate the evolutionary clock. Sarich (1977) proposes using the correlation between albumin immunological distance (AID), obtained by micro-complement fixation, and Nei's genetic distance. In diverse groups, an AID of 35 = 1.0 $\bar{D}$ units if 3/4 of the loci investigated are slowly evolving and the remainder rapidly evolving (Sarich 1977). Albumin data in many taxa (Sarich 1969, Gorman et al. 1971, Wallace et al. 1973) suggest that 10 AID units correspond to a divergence time of 6 million years and, thus, a genetic distance of 1.0 = 21 million years. Time estimates calculated by this method for divergence of *Enchenopa* host races are presented in Table 3-2. Divergence times range from approximately 3.5 myr for the *Juglans nigra* split from the ancestral stock to 0.25 myr for the most recent split between *C. scandens* and *Viburnum.* Nei (1972) advocates the relationship expressed by the formula $t = \bar{D}/(2cn_T\lambda a)$, where t is the period of time since a pair of populations or species become isolated, $\bar{D}$ the genetic distance (expected number of amino acid differences per protein), c the proportion of amino acid substitutions detected by electrophoresis, λa the rate of amino acid substitutions per polypeptide per site per year, and n_T the total number of codons (= number of amino acids) involved in the synthesis of a protein. Using the logic of Nevo et al. (1974), the above formula reduces to $t = 1.54 \times 10^6 \overline{D}$. This yields times of divergence that are exactly 13.63 times less than those calculated by the former method; these are presented in Table 3-2. While each method has certain advantages, we believe that it would be tenuous to attempt to choose between them. Therefore, they are included as upper and lower estimates of time of divergence for each *Enchenopa* host race. Both methods suggest that divergence has been a recent event and may be even more recent than the most conservative estimates. Regardless of these estimates, it is clear that genetic divergence of the *Enchenopa binotata* complex has occurred on these 6 broadly sympatric host plants in eastern North America.

6 The Role of Ants in the Life History and Dispersion of Treehoppers

We hypothesize that predaceous ants play a major role in affecting the survival and dispersion of treehoppers. Ants are attracted to and feed on the anal secretions (honeydew) of many Homoptera (Leston 1973a, 1973b, 1978). Similarly, we observed that aggregations of *Enchenopa* nymphs can be attended by ants. An examination of 150 branches with nymphs showed that 60 of these aggregations were attended by ants. The number of nymphs in attended aggregations averaged 3 times the number in non-ant-attended ones. These observations suggest that treehopper survival is greater where attending ants are present because they reduce predation from other invertebrates (Wood 1977, McEvoy 1979). Also, because ants are foraging for honeydew, large aggregations of treehoppers should be more attractive to ants than small ones.

To test this hypothesis 15 branches on each of 16 trees containing treehopper egg masses were tagged. When eggs hatched, counts of nymphs and ants were made until nymphs molted to adult (approx. 30 days later). As expected the mean number of nymphs/branch and the mean number of egg masses/branch were positively related. The mean number of nymphs on the 16 trees was positively related to the number of

Table 3-2. Time estimates of divergence for host races of the *Enchenopa binotata* complex. See text for additional explanation

Host race	Minimum estimate[a] (years B.P.)	Maximum estimate[b] (years B.P.)
J. nigra	256,000	3,486,000
P. trifoliata	188,000	2,562,000
R. pseudoacacia	116,000	1,575,000
C. canadensis	77,000	1,050,000
C. scandens	20,000	273,000

[a] Using the method of Nei (1972).
[b] Using the method of Sarich (1977).

aggregations on the tree which were attended by ants. Thus, the number of egg masses within a tree, the number of nymphs, and the number of tending ants were positively related. Nymphal survival on individual trees was related to the mean initial number of nymphs per branch and to the number of branches with nymphs that were ant-attended (Wood and Seilkop, unpubl.). When individual branches were considered independent of trees we found that branches with 11 or more nymphs were more consistently attended by ants and had a higher nymphal survival than those with few nymphs (Wood and Seilkop, unpubl.). Furthermore, McEvoy (1979) showed that treehopper survival is positively related to their proximity to ant nests.

We feel that ovipositional attractants in the egg froth, the clumped distribution of egg masses, and aggregation behavior in nymphs are all adaptations that concentrate the honeydew resource and encourage attendence by ants. Thus, as a consequence of ants, populations of *Enchenopa* tend to be scattered and disjunct.

7 Reproductive Isolation and the Evolution of Host Races

We propose that divergence of *Enchenopa* along host plant lines broadly fits a sympatric model for speciation. Based on our present knowledge of treehoppers, we postulate that the progenitor of North American *Enchenopa* was Neotropical and that females deposited eggs in masses covered with froth. *Enchenopa* occurs in the Neotropics on several different hosts (Ballou 1936), and Funkhouser (1951) proposes the tropics as the origin for treehoppers in general with subsequent radiations into north temperate regions. Egg froth in the tropics may serve as a protection from predators or parasites as well as a mechanism to ensure large aggregations of nymphs that concentrate honeydew and attract a defensive force of ants that are so abundant in the lowland tropics. Selection favoring females that were attracted to existing egg masses for oviposition should have been intense, given the fact that in our experiments nymphs on branches containing few egg masses attracted few ants and incurred severe mortality.

Enchenopa colonizing North America may have been polyphagous, or monophagous feeding on a single host with subsequent transfers to other hosts after its arrival. Regardless, inclement winters and synchronization of life history with the unique phenology of each host were problems to be incurred. Coordination of egg-hatch and nymphal development with optimal host nutrition (apparently when hosts are in flower) and avoidance of plant defenses were undoubtedly favored. As eggs hatched and nymphs developed in response to the phenology of their host plants, conditions which promoted host specialization and divergence were established. We see evidence for this today in the chronological differences in life histories of *Enchenopa* that occur on the various host plants. For instance, *Enchenopa* on *C. scandens*, *J. nigra* and *R. pseudoacacia* are effectively separated from each other by mating at different times during the season. On *Viburnum*, treehoppers mate in the evening, while those on *P. trifoliata* and *C. canadensis* are diurnal maters. *Enchenopa* on these latter two hosts disperse at different times during the season. Such differences should in part promote reproductive isolation among the host races.

Furthermore, tending ants appear to be a major factor in North America as well as the Neotropics, fashioning the life history of *Enchenopa* and dictating their highly clumped distribution. Successful colonization of new hosts should be a rare event considering the high mortality incurred by small populations of treehopper nymphs that fail to attract protective ants. Under these circumstances, dispersal should be as rare as we show, density-dependent, and occur only at high densities (see Wood and Guttman 1981). Consequently, as Janzen (1968) and Opler (1974) suggest, and Edmunds and Alstad show (1978; Chapter 2), host plants or individual trees may function as evolutionary islands with their isolated populations of coevolving herbivores.

Assortative mating within host plant species is almost assured considering disparate life histories, the mating system, and the insular nature of treehopper populations. We predict that the majority of females mate and deposit eggs either on the tree where they developed or on nearby conspecifics. The allelic heterogeneity we found among *Enchenopa* populations from adjacent conspecific host plants confirms this hypothesis (see also Guttman et al. 1981). This should promote genetic divergence of *Enchenopa* on different host plant species as well as among individual trees.

Selection of host-specific genotypes and assortative mating could lead to reproductive isolation among host races. The fact that females choose their developmental hosts for oviposition and mate mostly with males having developed on conspecific hosts suggests that significant divergence has occurred. The genetic differences we present among entities in the *Enchenopa* complex indicate that sufficient time has elapsed for the development of post-mating isolation mechanisms. Treehoppers on *J. nigra, P. trifoliata, C. canadensis* and *R. pseudoacacia* are electrophoretically the most distinct. Offspring from mixed matings among these *Enchenopa* or between these and treehoppers on *Viburnum* or *C. scandens* may have reduced fitness. Although *Enchenopa* on the latter two hosts are closest genetically, they are ecologically well-separated by differences in mating phenology. In this proposed model we feel that slight life history shifts of *Enchenopa* in response to host plants combined with behavioral and ecological factors are all that are needed to produce reproductively isolated species.

Further tests of our model proposed by Wood (1980) depend on:

(1) Detailed studies of tropical *Enchenopa,* since evidence for a tropical progenitor is tenuous.
(2) Understanding the mechanisms that coordinate treehopper life histories with host phenology.
(3) Determining the relative contributions of genetics and environment (host plants) to the variation in *Enchenopa* life histories.

Summary. Given the life history, host-plant interactions, and behavior, we feel that *Enchenopa* has diverged along host plant lines into a complex of reproductively isolated populations. The mechanism suggested for this divergence broadly fits into a sympatric model.

Acknowledgments. We appreciate the efforts of both the editors of this volume and an anonymous reviewer for their help in clarifying the manuscript. Portions of this work were supported by National Science Foundation Grants BMS 74-19764 A01, and a Miami University Faculty Research Grant.

Published with the approval of the Director of the Delaware Agricultural Experiment Station as Miscellaneous Paper No. 918, Contribution No. 495, of the Department of Entomology and Applied Ecology, University of Delaware, Newark, Delaware 19711, U.S.A.

References

Alexander, R. D.: Life cycle origins, speciation and related phenomena in crickets. Q. Rev. Biol. 43, 1-41 (1968).

Alexander, R. D., Bigelow, R. S.: Allochronic speciation in field crickets and a new species, *Achetis veletis*. Evolution 14, 334-346 (1960).

Ballou, C. H.: Insect notes from Costa Rica in 1935. Insect Pest Surv. Bull. 16, 454-479 (1936).

Bush, G. L.: Sympatric host race formation and speciation in frugivorous flies of the genus *Rhagoletis* (Diptera: Tephritidae). Evolution 23, 237-251 (1969).

Bush, G. L.: Modes of animal speciation. Ann. Rev. Ecol. Syst. 6, 339-364 (1975).

Dixon, A. F. G.: Applied ecology: Life cycles, polymorphism and population regulation. Ann. Rev. Ecol. Syst. 8, 329-353 (1977).

Edmunds, G. F., Alstad, D. N.: Co-evolution of insect herbivores and conifers. Science 199, 941-945 (1978).

Funkhouser, W. D.: Homoptera Fam. Membracidae. Genera Insectorum 208, 1-383 (1951).

Gorman, G. C., Wilson, A. C., Nakanishi, M.: A biochemical approach towards the study of reptilian phylogeny: Evolution of serum albumin and lactic dehydrogenase. Syst. Zool. 20, 167-185 (1971).

Gorman, G. C., Soule, M., Yang, S. Y., Nevo, E.: Evolutionary genetics of insular Adriatic lizards. Evolution 29, 52-71 (1975).

Guttman, S. I., Wood, T. K., Karlin, A.: Genetic differentiation along host plant lines in the sympatric *Enchenopa binotata* Say Complex. (Homoptera: Membracidae). Evolution (1981) (in press).

Huettel, M. D., Bush, G. L.: The genetics of host selection and its bearing on sympatric speciation in *Procecidochares* (Diptera: Tephritidae). Entomol. Exp. Appl. 15, 465-480 (1972).

Janzen, D. H.: Host plants as islands in evolutionary and contemporary time. Am. Nat. 102, 592-595 (1968).

Kim, Y. J., Gorman, G. C., Papenfuss, T., Roychoudury, A. K.: Genetic relationships and genetic variation in the amphisbaenian genus *Bipes*. Copeia 1976, 120-124 (1976).

Knerer, G., Atwood, C. E.: Diprionid sawflies: Polymorphism and speciation. Science 179, 1090-1099 (1973).

Leston, D.: Ecological consequences of the tropical ant mosaic. Proc. VII. Congr. IUSSI, London, 1973a, pp. 235-242.

Leston, D.: The ant mosaic–tropical tree crops and the limiting of pests and diseases. PANS 19, 311-341 (1973b).

Leston, D.: A neotropical ant mosaic. Ann. Entomol. Soc. Am. 71, 649-653 (1978).
Maynard Smith, J.: Sympatric speciation. Am. Nat. 100, 637-650 (1966).
Mayr, E.: Populations, Species and Evolution. Cambridge: Harvard Univ. Press, 1970.
McEvoy, P. B.: Advantages and disadvantages to group living in treehoppers (Homoptera: Membracidae). Misc. Publ. Entomol. Soc. Amer. 11, 1-13 (1979).
Metcalf, Z. P., Wade, V.: General catalogue of Homoptera: Membracidae. Fas. 1 and 2. Raleigh, N.C.: North Carolina State Univ. Press, 1965.
Nei, M.: Genetic distance between populations. Am. Nat. 106, 283-292 (1972).
Nevo, E., Kim, Y. J., Shaw, C. R., Thaeler, C. S., Jr.: Genetic variation, selection, and speciation in *Thomomys talpoides* pocket gophers. Evolution 28, 1-23 (1974).
Opler, P. A.: Oaks as evolutionary islands for leaf mining insects. Am. Sci. 62, 67-73 (1974).
Sarich, V. M.: Pinniped origins and the rate of evolution of carnivore albumins. Syst. Zool. 18, 286-295 (1969).
Sarich, V. M.: Rates, sample sizes, and the neutrality hypothesis for electrophoresis in evolutionary studies. Nature (London) 265, 24-28 (1977).
Simberloff, D.: Species turnover and equilibrium island biogeography. Science 194, 572-578 (1976).
Tauber, C. A., Tauber, M. J.: Sympatric speciation based on allelic changes at three loci: evidence from natural populations in two habitats. Science 197, 1298-1299 (1977a).
Tauber, C. A., Tauber, M. J.: A genetic model for sympatric speciation through habitat diversification and seasonal isolation. Nature (London) 268, 702-705 (1977b).
Tauber, C. A., Tauber, M. J., Nechols, J. R.: Two genes control seasonal isolation in sibling species. Science 197, 592-593 (1977).
Wallace, D. G., King, M. L., Wilson, A. C.: Albumin differences among ranid frogs: Taxonomic and phylogenetic implications. Syst. Zool. 22, 1-13 (1973).
Wood, T. K.: The role of parent females and attendant ants in the maturation of the treehopper, *Entylia bactriana* (Homoptera: Membracidae). Sociobiology 2, 257-272 (1977).
Wood, T. K.: Divergence in the *Enchenopa binotata* Say complex (Homoptera: Membracidae) Effected by host plant adaptation. Evolution 34, 147-160 (1980).
Wood, T. K., Guttman, S. I.: The ecological and behavioral basis for reproductive isolation in the sympatric, *Enchenopa binotata* complex. Evolution (1981) (in press).
Yang, S. Y., Soule, M., Gorman, G. C.: *Anolis* lizards of the eastern Caribbean: A case study in evolution. I. Genetic relationships, phylogeny and colonization sequence of the *roquet* group. Syst. Zool. 23, 387-399 (1974).

Part Two

Population and Species Variation in Life Histories

For life histories, as for other combinations of traits, variation and covariation provide the raw material upon which natural selection acts to guide the course of evolution. The development of predictive theory therefore requires as its first step consideration of the variation present in natural populations. Such consideration, and its consequences for both basic and applied aspects of life history analysis, is the subject of this section. Variation is surveyed and analyzed at the level of populations, where it may yield insight into the microevolutionary processes directing the formation of current adaptations, and of species, where adaptations are products of previous evolutionary history. Insights deriving from a comparative analytical approach to life histories are apparent throughout.

In Chapters 4 and 5, H. Dingle and W. S. Blau examine life history variation in populations of the same species in insects that occur at both tropical and temperate latitudes. Dingle compares three populations of the milkweed bug, *Oncopeltus fasciatus,* in Iowa, Florida and Puerto Rico. The former is a diapausing long-distance migrant, while the latter two move locally in habitats and climatic regimes which differ considerably in predictability for the respective bug populations. Behavior thus plays an important role in life history evolution and in fact influences such population parameters as "r_{max}." Other factors, such as differences in the partitioning of eggs among clutches (25 eggs every day vs. 50 eggs every other day), may also be of little significance to r_{max} but, in combination with choice of oviposition sites (a behavioral trait), they have profound influences on survivorship in the field. Blau examines life history differences between a bivoltine diapausing temperate population of the butterfly *Papilio polyxenes* and a continuously breeding tropical population. Classical predictions of life history theory concerning life history differences between tropical and temperate forms are of minimal utility for several reasons, including the fact that for this species the tropical habitat may be more uncertain in its temporal variability than the temperate one. Flight behavior and colonizing ability are important elements of the population biology. Differences in body size arising for physiological reasons may also influence life histories.

Chapters 6 and 7, by W. B. Showers and C. A. Istock, deal with the genetic variance underlying life history characters. Showers considers diapause variation in the European corn borer, a particularly interesting species because of its relatively recent introduction into North America. Geographic variation in diapause and voltinism reveals that evolution of these traits in response to local conditions has already taken place. Selection experiments suggest that additive genetic variance is a significant contributor to population differences observed. Since the corn borer is an important pest species, the presence of this genetic variance could be of major significance in the development of pest management strategies and techniques. Istock uses an interesting cumulative definition of fitness to analyze the genetic basis of life history adaptations, especially the significance of local genetic variance in diapause in the pitcher plant mosquito, *Wyeomyia smithii.* He discusses a theoretical model for the maintenance of genetic variance via "fluctuating-stabilizing" selection and demonstrates, through crossing of latitudinally disparate stocks, that gene differences influencing life histories do indeed occur between geographic populations. As with the corn borer, the results, especially as they concern genetic response to variable selection, may have implications for pest management.

Finally, Chapter 8, by D. W. Tallamy and R. F. Denno, focuses on differences among species. The insects whose life histories are considered are lacebugs (Tingidae) exploiting woodland and early successional plants. After examining current life history theory, the authors conclude that it is inadequate to predict the evolution of reproductive patterns in these bugs. The reason is that different combinations of traits, including aspects of behavior usually not considered in formulating theory, may be equally adaptive in particular environments. Phylogenetic and evolutionary histories may provide distinct arrays of characters and degrees of variation upon which current selective pressures may act. The result can be diverse covarying combinations of characters achieving the same adaptive end.

Chapter 4

Geographic Variation and Behavioral Flexibility in Milkweed Bug Life Histories

HUGH DINGLE

1 Introduction: Behavior and Life Histories

The evolution of life histories is a topic of great interest among evolutionary ecologists. The reasons are clear. Patterns of survival and of production of young determine the number of descendants in succeeding generations and hence fitness. Traditionally investigators have used mean values of life table statistics to investigate life history strategies, and indeed much current theory is based on the use of means and on assumptions that constant percentages of individuals breed even in fluctuating environments (reviewed in Stearns 1976). An example of the theory so generated is the much discussed concept of r- and K-selection. Implicit in such theories is the underlying notion that life histories represent some optimal genotype; variation and life history flexibility, whether genetically or environmentally influenced, have been largely ignored (see Nichols et al. 1976 for additional discussion).

It is becoming increasingly apparent, however, that such variation and flexibility are critically important in life history evolution. The choice of where and when to breed, for example, and the ability to move to breeding sites (Taylor and Taylor 1977, 1978) are as important as such traditional life table statistics as clutch or litter size or age at first reproduction. Indeed these latter variables may be determined behaviorally as a consequence of the choices made by the organism (e.g. Dingle 1968). Life history flexibility in several vertebrates including amphibians, birds and mammals is reviewed by Nichols et al. (1976). Flexible breeding responses as a consequence of environmental unpredictibility are especially well-developed in desert species, which may display rapid physiological and behavioral changes to reproduce whenever and wherever rainfall permits. Behaviorally modified phenologies and their life history consequences are, if anything, even more prevalent in insects.

Two primary behaviors which influence the evolution of insect life histories are migration and diapause. That these two responses cannot be considered separately from the life histories of which they are integral parts was indicated clearly by J. S. Kennedy (1961). Kennedy pointed out that, although on the face of it migration and diapause looked like alternatives, the former taking the insect to a new habitat while the latter allowed it to remain in the old, they were in fact physiologically similar in many ways. Both result in suspension of growth, development and/or reproduction. The role they play in the life history is thus obvious, as they allow the insect to choose between reproducing "here and now," "here and later," or "elsewhere" in Solbreck's (1978) terminology. The evolution of appropriate interactions between migration, diapause and reproduction in the shaping of life history strategies is now apparent in a number of insects (Dingle 1974a, 1979, Kennedy 1975, Solbreck 1978, Vepsäläinen 1978). And, since a number of economically important species migrate, enter diapause, or both, an understanding of the coadaptation of these behaviors with demographic traits is of more than simply esoteric interest. Understanding the spatial and temporal constraints of varying selective regimes is crucial not only to the development of relevant theory but also to the design of management techniques (papers in Rabb and Kennedy 1979).

In our laboratory we are interested in the role of migration and diapause in the evolution of life histories in milkweed bugs (*Oncopeltus* spp.). Our studies involve comparisons both within and among species (Dingle 1978, Dingle et al. 1980a, 1980b), but I shall concentrate here on a comparison among three populations of *O. fasciatus,* the species which appears to have the most widespread distribution (O'Rourke 1977). Within-species comparisons are most likely to reveal available evolutionary alternatives and to minimize differences resulting not from current selection but from basic ground plans rooted in the evolutionary past (Solbreck 1978). Such ground plans are often not easily altered by selection. The populations compared are from Iowa, Florida and Puerto Rico, chosen to represent disparate selective regimes. I shall first discuss the natural history of *O. fasciatus* in the three locations, then summarize studies on diapause, migration, and life table statistics. Finally, I shall attempt to relate the observed differences in the life history patterns to the selective regimes encountered by the populations, indicating the importance of behavior.

2 Geographic Variation in the Natural History of *Oncopeltus fasciatus*

In Iowa *O. fasciatus* is a seasonal migrant, with most reproduction and population growth taking place on milkweed plants (mostly *Asclepias syriaca*) in late summer (Fig. 4-1). The first adults appear in late June or early July, as indicated in the lower panels in Fig. 4-1. These incoming migrants mate and reproduce so that the initial first-instar nymphs appear 2-3 weeks later. The population grows rapidly as more migrants arrive and lay eggs, and the nymphs mature to produce the first adult descendants in late August or September, depending on temperature conditions. The difference between a cool summer and a normal summer is indicated in Fig. 4-1, which contrasts 1967, the coolest summer of the century, with daily temperatures usually 3-4 °C below the previous 30-year average, and 1968, with daily temperatures close to the average. The delay in the emergence of the summer generation of adults in 1967 is obvious.

The eclosing adults of late summer emerge to rapidly decreasing daylengths, and these short days induce an adult reproductive diapause (Dingle 1974b). These diapausing adults emigrate, resulting in the collapse of the population in late September and

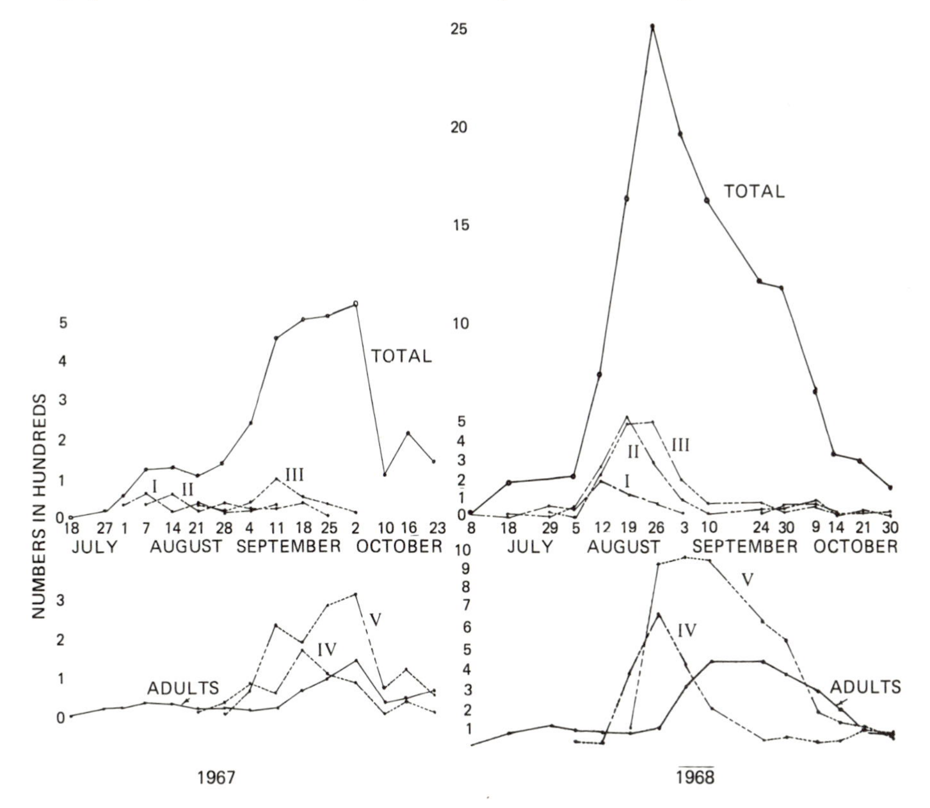

Figure 4-1. Population growth in Iowa populations of *Oncopeltus fasciatus* censused on *Asclepias syriaca* in the summers of 1967 and 1968. Growth was considerably less in the cool summer of 1967. Note that immigrant adults arrive in July; the decline in the population in the fall results from departure of newly eclosed adults. Roman numerals indicate changes in numbers of the various instars.

October. Field evidence for diapause includes the observation that after the middle of September no mating pairs are recorded in censuses. In laboratory tests of flight durations using tethered insects (discussed below), it was found that bugs in diapause displayed flights of long duration, presumably migratory, during the diapause period (Dingle 1978).

Because of the late summer diapause, Iowa populations of *O. fasciatus* are essentially univoltine. At some sites in some years, however, there may be a small second generation. Close inspection of the growth curves for 1968 (Fig. 4-1), for example, reveals a slight increase in the number of early instars in late September and early October. These represent a small second generation, a few of which reached adulthood before the severe frosts of November killed all remaining nymphs. In temperate areas with a somewhat longer growing season, a nearly complete second generation may emerge at some sites. Such was the case in the Maryland populations studied by Ralph (1977).

The local pattern of movement in *O. fasciatus* at a site near Iowa City, Iowa, was examined with a mark-recapture study. During three periods spanning July through

September (see Table 4-1), bugs in an area approximately two hectares in extent were marked on the wings with a felt-tip marker (a detailed analysis including a description of the study site will be published elsewhere). The population was censused in each case one week after the marking episode, and the number of marked bugs noted. The results are presented in Table 4-1, and indicate that bugs arriving in midsummer were more likely to be recaptured than bugs emerging in mid- to late September. Of 250 bugs marked in July and August, 55 were sighted again, while in September only 25 of 271 were recaptured. I interpret this to mean that a high proportion of immigrant bugs remain at the site to reproduce, while their offspring emigrate in the autumn following maturation. Additional evidence for the former conclusion comes from the fact that some of the July-August bugs, which had been marked with individually identifying codes, were sighted repeatedly for up to 6 weeks.

An experiment in which some individuals were wing-clipped also demonstrates that emigration is a significant contributor to the disappearance of adults in September. The posterior half of both pairs of wings was removed from approximately half the bugs in late September. The assumption was that, if higher mortality accounted for the recovery of fewer September bugs, the recapture rate of both intact and wingless bugs would be the same. If, on the other hand, the chief cause was emigration, then fewer intact bugs would be recovered. The latter turned out to be the case, as is indicated in Table 4-1. Over twice as many wing-clipped bugs were recaptured, strongly suggesting that emigration is largely responsible for the disappearance of adults in the autumn. The recovery of a few intact bugs was expected because these bugs fly post-tenerally (Dingle 1965), and a few would not have completed the teneral period prior to resighting.

The annual cycle of *O. fasciatus* in southern Florida is quite different from that in Iowa. A full study is reported by Klausner (1979), so a brief summary will suffice here. The chief milkweed host plants are *Asclepias curassavica* and *Sarcostemma clausa,* the latter a vine growing along the edges of swamps and sloughs. In addition there is some reproduction and population growth on oleander bushes (*Nerium oleander*), although both are reduced on oleander relative to milkweeds (Klausner et al. 1980). Two sites, a cow pasture dominated by *A. curassavica* and an area of swamps and ponds with extensive stands of *S. clausa,* were sampled at intervals throughout 1978 by counting milkweed bugs along a 100-meter transect at each site.

The results of the transect sampling, summarized in Fig. 4-2, indicate that *O. fasciatus* populations are influenced by the phenology of the host plants. Reproduction

Table 4-1. Results of mark-recapture experiments, from the Iowa population of *Oncopeltus fasciatus* in 1979

	Lost	Recovered at 7 days
July-August	195	55
Mid-September	148	15
Late September	98	10
Wing clipped	82	22

Chi-square (time) = 16.33, $P < 0.005$, 2 df.
Chi-square (clipped) = 5.85, $P < 0.025$, 1 df.

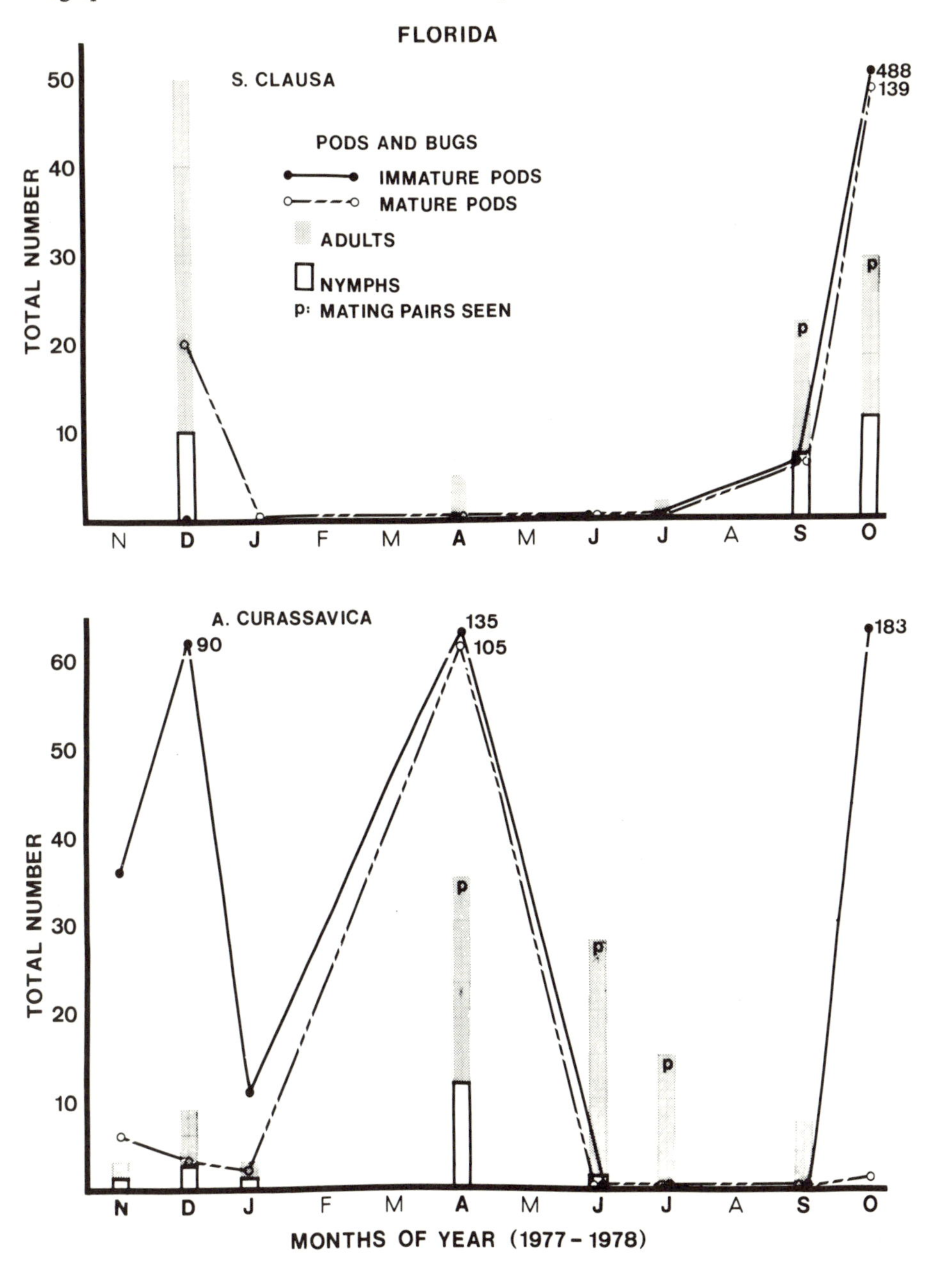

Figure 4-2. Population changes during one year in Florida *O. fasciatus* on two host milkweeds, *Sarcostemma clausa* and *Asclepias curassavica*. Reproduction is associated with fruiting host plants, and is concentrated in the late spring and early summer on *A. curassavica* and in the autumn on *S. clausa*. Considerable variation is present among years and sites. (Data from Klausner 1979.)

in the bugs, as indicated by the presence of nymphs, occurs only when mature milkweed pods are available. Thus population growth occurs primarily in the spring (April-June) on *A. curassavica* and in autumn and early winter (September-December) on *S. clausa.* A few adults were present at every census, but evidently no offspring were produced in the summer, even though mating pairs were seen at the *A. curassavica* site. Oleander is a host plant in the summer with production of offspring, and these mating pairs may have immigrated from nearby oleander bushes. Our data suggest that *O. fasciatus* in southern Florida probably breeds throughout most of the year, but with no or little reproduction in the late winter dry season (January-February) and late summer wet season (August).

Superimposed on this general pattern is variation generated by environmental uncertainty. Two examples will illustrate this point. First, January 1978 was exceptionally cold for the region, and the absence of breeding (the nymphs observed on *A. curassavica* were fifth instars) may have been a consequence of low temperatures. Laboratory results indicate that *O. fasciatus* would have no difficulty breeding in some years at daytime temperatures (20-23 °C) normal for this part of Florida in January. Secondly, seasonal rainfall can be even more erratic. A failure of the rains in April-May 1976 resulted in no flowering or seed-set in *A. curassavica.* In April, a time when there were many nymphs and reproducing adults in 1978, there were no bugs at all found in 1976. The climate and host plant phenology of southern Florida thus permit almost continuous breeding in most years, albeit with some seasonal fluctuation, but with unpredictable variation among years, resulting in periods when no reproduction is possible.

Finally, in Puerto Rico the climate is equable throughout the year, with relatively little temperature or rainfall variation (Dingle et al. 1980a). As would be expected, flowering and seed-set in the primary host plant, *A curassavica,* is continuous, as is reproduction in *O. fasciatus.* There is, however, considerable variation in host plant phenology among sites, as a function of human activity and other factors, with consequent influence on milkweed bug populations. This is illustrated in Fig. 4-3, which shows censuses at two sites, a few hundred meters apart, during October and early November, 1979. Clearly, population growth was not synchronized in the two sampling areas, but in each it was correlated with the availability of milkweek pods. Immigration into a site with maturing pods suggests considerable local movement by the bugs.

The relatively ephemeral nature of specific *A. curassavica* patches is also indicated by variation among years. Three census areas which had populations of 971, 214 and 0 bugs in early June, 1978, had populations of 14, 28 and 84 respectively at the same time in 1979. High bug densities coincided with high frequencies of mature pods on the *A. curassavica* plants, and vice versa. The natural history of *O. fasciatus* in Puerto Rico evidently involves local movements among asynchronously fruiting milkweed patches. A similar finding for other species of *Oncopeltus* has been reported by Root and Chaplin (1976) for sites in Colombia. A detailed study of *O. fasciatus* in Puerto Rico will appear elsewhere (W. S. Blau, in prep.).

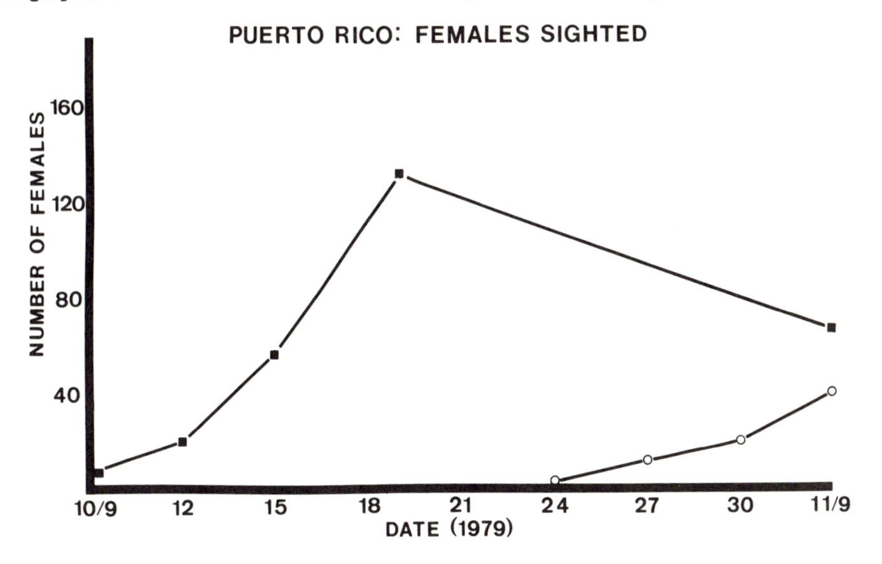

Figure 4-3. Population changes in Puerto Rico *O. fasciatus* in two neighboring patches of *A. curassavica*. Population increase accompanies fruiting by the host plant which is asynchronous across patches. (Data from W. S. Blau, unpublished.)

3 Diapause

Diapause in the *O. fasciatus* populations studied is measured by the duration of the interval between adult eclosion and the age at first oviposition. The distribution of ages at first reproduction so measured is continuous, but we have operationally defined diapause as any delay in oviposition beyond 30 days posteclosion, as temperate bugs on long days have all produced eggs by this time (Dingle et al. 1980a). The diapause response in the three geographic populations discussed here was determined at two photoperiods, LD 14:10 and LD 11:13 at each of two temperatures, 23 and 27 °C, in a 2 X 2 factorial design. The photoperiods were chosen to represent approximately the extremes encountered by *O. fasciatus* over most of its range. The two test temperatures were selected because reproduction occurs in both, because both are likely to be encountered at a number of locations over the species distribution, and because Iowa bugs display diapause at 23 °C under short days but not at 27 °C.

The results of the diapause experiments are indicated in the interaction plots displayed in Fig. 4-4. All four regimens are given for each population, with mean ages at first reproduction (ovipostion) and their standard errors plotted above each temperature. The two points for each photoperiod are then connected by lines. Consider first the results for Iowa shown by the heavy lines. At 27 °C there is little difference between points, indicating at most only a slight influence of photoperiod. In proceeding from 27 to 23 °C, the lines for the two photoperiods diverge markedly. On short days at the lower temperature, individuals enter diapause, as indicated by a mean age of first reproduction of 67.5 days. On long days there is an expected delay in oviposition due to the lower temperature, but no diapause. The strong divergence of the photo-

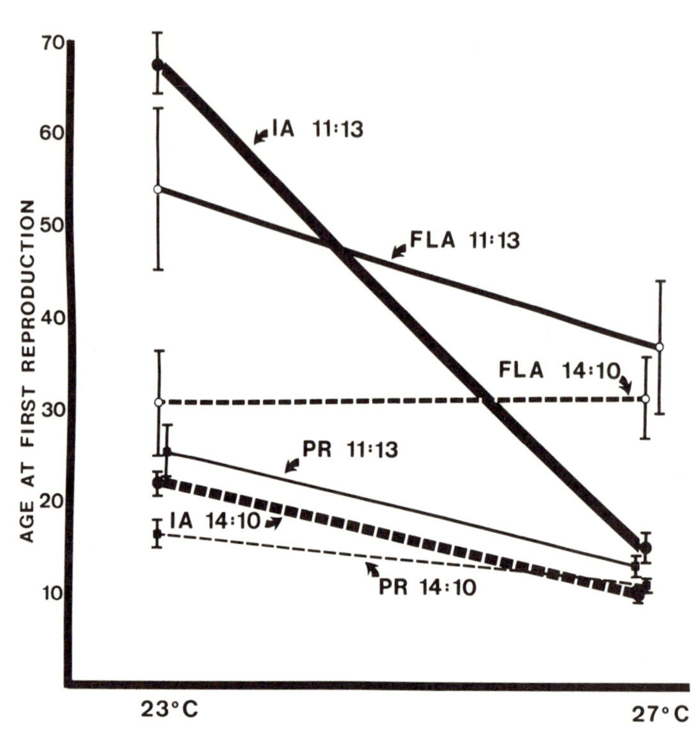

Figure 4-4. Interaction plots showing responses of the three *O. fasciatus* populations to the four photoperiod-temperature regimens. A difference between points on the same line indicates a response to temperature (e.g., at LD 11:13 Florida bugs delay reproduction to about 54 days at 23 °C as opposed to about 39 days at 27 °C), while a difference between lines indicates a response to photoperiod (note, e.g., Florida). Diverging lines indicate an interaction between temperature and photoperiod, as is particularly evident in Iowa (IA) and Florida (FLA), where delayed reproduction (diapause) occurs at low temperatures especially in short days. Age at first reproduction is in days posteclosion. PR, Puerto Rico.

period lines indicates an interaction effect between photoperiod and temperature, i.e., Iowa bugs diapause in short days, but only at the lower temperature.

An interaction effect is also apparent in the Florida bugs, with a conspicuous delay in the onset of reproduction apparent in short day and low temperature. Two other aspects of the responses of the Florida population should also be noted. First, under all conditions some portion of the population diapauses, as evidenced by mean ages of first reproduction greater than 30 days. This contrasts with Iowa, where diapause occurred only at LD 11:13 and 23 °C. Secondly, the variances for Florida bugs were much greater than for either of the other two geographic origins. This reflected the fact that a full range of responses was found in Florida individuals under all conditions (Fig. 4-5 and Dingle et al. 1980a). Any delay in reproductive onset resulting from lower temperatures was obscured by the high variances at both 23 and 27 °C; as indicated by the horizontal LD 14:10 line for Florida, no temperature difference was apparent.

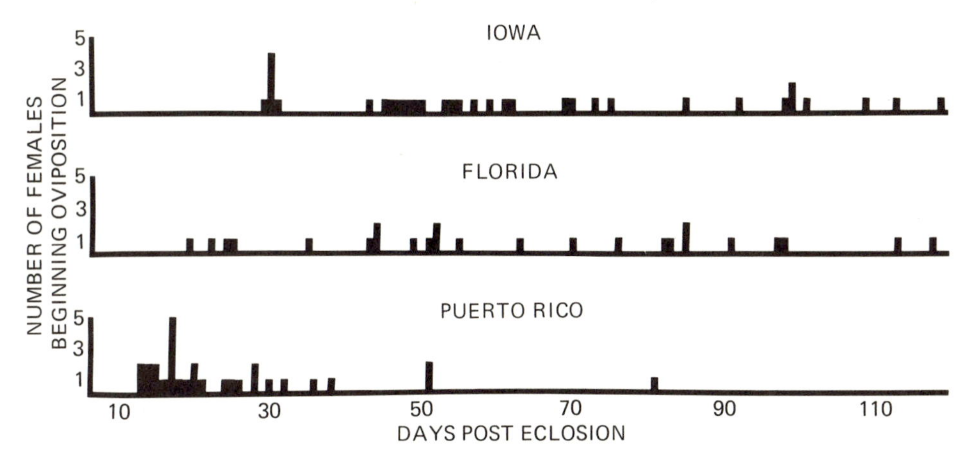

Figure 4-5. Frequency histogram showing distribution of responses of the three *O. fasciatus* populations to the short-day, low-temperature regimen. Note especially the variability in Iowa and Florida.

Finally, diapause, as determined by our operational 30-day criterion, is not apparent in the mean response of the *O. fasciatus* population from Puerto Rico. As expected, there was a slight delay in reproduction at the lower temperature. Day length was also not totally without effect, as indicated by the separation of the two Puerto Rico photoperiod lines. There was a small but consistent delay in age at first reproduction in short days at both temperatures. At present we do not know whether this delay represents some aspect of local adaptation in Puerto Rico or occasional immigration of some diapause genotypes from other *O. fasciatus* populations to the West.

To further indicate the amount of variability associated with developmental responses to temperature and photoperiod in *O. fasciatus,* interval histograms of the results for the three populations at LD 11:13 and 23 °C are shown in Fig. 4-5. These were the most variable results for Iowa and Puerto Rico bugs, which displayed much narrower ranges under the other three regimens. Note that even in the Puerto Rico population a few bugs do diapause, the individuals initiating oviposition later than 30 days posteclosion being largely responsible for the slight interaction effect between temperature and photoperiod seen for Puerto Rico in Fig. 4-4. Note also the high variability in the Florida bugs; similar variability was found under all conditions in this population (Dingle et al. 1980a), and is apparently an important characteristic of these bugs. Its importance is addressed further in the Discussion section below.

4 Flight

We index migration in *Oncopeltus* by determining duration of tethered flights (Dingle 1978, 1979, Dingle et al. 1980b). The bugs are glued to a stick at the prothorax and lifted free of the substrate. In this circumstance the insects usually fly readily, often for several hours, although sometimes wind on the head is necessary initially to stimulate flight. That tethered flight in fact indexes migration is suggested because (1) immigrant bugs in early summer in Iowa, (2) Iowa bugs in the autumn just prior to emigration, and (3) bugs entering a milkweed patch from which all individuals have been previously removed all display increased flight duration relative to controls (Caldwell 1969, 1974, Blakley 1977). These bugs are likely to be migrants; similar results for other insects are reported by a number of workers.

The flight tests discussed here were conducted on bugs which were reared at LD 14:10 23 °C. Migratory flights for *O. fasciatus* from north central North America have been previously recorded under these conditions (Dingle 1965 et seq.), indicating that this regimen is appropriate for assessing migratory flight. A temperature of 27 °C suppresses flight in these migrants (Dingle 1978). The long-day photoperiod also does not induce diapause in Iowa or Puerto Rico bugs, an important consideration since diapause can have a profound influence on flight (Dingle 1978, 1979). As indicated above, however, a significant proportion of the Florida sample does diapause under these conditions. To control for this effect, comparisons of flight are made between bugs, 8, 10 and 12 days posteclosion only. Bugs at this age are postteneral, necessary for flight to occur, and prereproductive, since reproduction, which suppresses flight, begins at around 15 days at this temperature in individuals not entering diapause. All bugs tested should thus have been physiologically similar with respect to reproduction and diapause.

The results of the flight tests are presented in Table 4-2. For each population the number of bugs flying in each of four categories from no flight to flights of 30 minutes or more are indicated. The durations recorded represent the sums of 5 consecutive flights for each bug and, since individuals were tested 3 times (at 8, 10 and 12 days posteclosion), only the largest of the daily sums are included. As indicated by the proportion flying for over 30 minutes, most long-duration flights occurred in the Iowa population, a not unexpected result in view of the known migratory tendencies of these temperate zone bugs. Most of these > 30-minute flights in fact lasted for an hour or more. The least long duration flight was displayed by the Puerto Rico sample, in which 6 out of 62 also failed to fly at all. Differences between any pair of samples were significant ($P < 0.02$ or better). Other northern samples are similar to Iowa in their flight performances, and other tropical samples resemble Puerto Rico in showing little long-duration flight (Dingle et al. 1980b). The conclusion from these results is that there is some migration in the Florida population, but very little in the bugs from Puerto Rico. The differences in flight performance among temperate and tropical samples of *O. fasciatus* (and among other species of *Oncopeltus*) are discussed in detail in Dingle et al. (1980b).

Table 4-2. Comparison of tethered flight performance for three populations of *Oncopeltus fasciatus*

Population	No flight	Number flying for: <5 min	5-30 min	>30 min	Total	Proportion flying >30 min
Iowa	0	114	27	34	185	0.238
Florida	1	30	6	8	45	0.178
Puerto Rico	6	48	5	3	62	0.048

5 Life Table Statistics

Migration in insects is usually associated with colonization of new habitats, and there is an extensive literature on the relation between colonization and life history strategies (reviewed by Lewontin 1965, Dingle 1972, 1974, Stearns 1976). In summary, the argument runs as follows: Invasion of a new habitat usually means entering an essentially empty universe as far as the colonizing species is concerned. Under these conditions selection favors those individuals that can produce descendants most rapidly and in the greatest numbers. The result is a life history pattern of early reproduction and high fecundity. This suite of characters (including others which need not concern us here) has been termed an "r-strategy" because the life table statistics in question lead to high values of r, the so-called intrinsic rate of increase (MacArthur and Wilson 1967). Because determinations of r require certain assumptions, including a stable age distribution, they are not entirely realistic when applied to unstable populations of migrant colonizers. They do, however, provide a useful index of potential for increase and do not deviate much from other measures with fewer assumptions (e.g., Laughlin 1965, Solbreck 1978).

Some values of r for various species of insect with high and low levels of migratory behavior and for some populations within species are presented in Table 4-3. Comparisons among species do in fact suggest that higher r's occur in migrants. The large African *Dysdercus fasciatus* and Central American *D. bimaculatus* display conspicuously higher values relative to their respective congeners *D. superstitiosus* and *D. mimulus.* Similar differences are found in *Oncopeltus* and *Tribolium.*

Table 4-3. Values of r (intrinsic rate of increase day^{-1}) for various species of insects and for populations within species, indicating differences among taxa with high and low migratory tendencies

Species	r (migratory)	r (low migratory)
Oncopeltus fasciatus[a,b]	0.037-0.044	0.050
O. unifasciatellus[b]	–	0.034
Dysdercus fasciatus[c]	0.094	–
D. bimaculatus	0.106	–
D. superstitiosus	–	0.062
D. mimulus	–	0.066
Tribolium castaneum[d]	0.128	–
T. confusum	–	0.100
Neacoryphus bicrucis[e]	0.109-0.116	0.123
Aphis fabae[f]	0.91-1.27	0.99-1.52

[a] Dingle (1968).
[b] Landahl and Root (1969).
[c] J. A. Derr, B. M. Alden and H. Dingle (MS in prep.).
[d] Ziegler (1976).
[e] Solbreck (1978).
[f] Dixon and Wratten (1971).

On the other hand, an apparent paradox emerges if we compare r for high and low migratory populations of the same species. In these cases r is lower in the migrants. The paradox is more apparent than real, however, if we take behavior into account (see also Kennedy 1975). If current habitats are deteriorating, insects must choose between reproducing immediately or moving to a new location with resultant reproductive delay. Clearly with marked habitat deterioration, selection favors movement even if it results in delayed reproduction, reduced fecundity, and hence reduced r (Dingle 1968). In the three cases presented in Table 4-3, *O. fasciatus* (Iowa) diapauses and migrates on short days to escape winter, *Neacoryphus bicrucis* does so under similar conditions or when food is short, and *Aphis fabae* produces alates when crowded or when the host plant senesces. Alterations in life histories thus come about in response to adaptive behavior.

A close examination of life table statistics for the three sample populations of *O. fasciatus* discussed here reveal such a life history alteration in the Florida bugs. Table 4-4 allows comparisons among the populations for some traits in insects maintained in an LD 14:10 27 °C regimen. This regimen was chosen because it both minimized diapause, which occurred only in a portion of Florida individuals, and presented conditions likely to be representative of those encountered, at least in the temperate summer, by bugs at all three geographic locations. Traits influencing potential population growth (and hence the rate of production of descendants) are quite similar in the Iowa and Puerto Rico groups. Thus there are only slight and not statistically significant differences between these bugs in age at first reproduction, total fecundity, life span, r, the replacement rate R_0, and the doubling time of the populations. The contrast shows up with Florida. Since a proportion of this sample diapauses, there is a consequent delay in mean age at first reproduction to 20.9 days instead of the approximately 11 days observed in the other two groups. Delays in reproduction meant that some females died before ovipositing or before reaching maximum egg output, in spite of the overall increase in life span, with resultant reduction of mean fecundity. Later age at first reproduction and lower fecundity combine to reduce values of the various statistics describing population growth. Note that it is diapause, a behavioral response to adverse conditions, which exerts a profound influence on life table statistics.

There is an interesting life table difference between Iowa and Puerto Rico which, although it exerts no influence on population parameters, implies that natural selection has been acting differentially. Iowa bugs produce more clutches with about half the

Table 4-4. Comparison of selected life table statistics for three populations of *Oncopeltus fasciatus* at LD 14:10 27 °C

Statistic	Iowa	Puerto Rico	Florida
Wing length	12.56	10.95	13.17
Age at first reproduction (days)	11.0	11.2	20.9
Number of clutches	32.9	19.5	17.3
Total fecundity	849	866	570
Clutch size	25.7	47.2	30.5
Age at death (days)	49.5	44.5	54.1
r (day^{-1})	0.121	0.126	0.098
R_0	283	306	136
Doubling time (days)	5.7	5.5	7.1

clutch size of Puerto Rico. Examination of oviposition patterns reveals that Iowa females lay approximately 25 eggs every day while Puerto Rico females lay roughly 50 eggs every other day. Field studies revealed the probable reason for this difference.

In conjunction with a comparative field study of juvenile mortality in *O. fasciatus* which will be published elsewhere (W. S. Blau, in prep.), survival to hatching was determined at sites in Iowa and Puerto Rico for clutches of 25 and 50 eggs. These clutches were constituted on the day of oviposition from eggs produced by laboratory stocks. Eggs in each "clutch" were selected from mothers collected previously at the site in question. These clutches were then placed out on milkweed plants in the field in locations such as spider and caterpillar webs and rolled leaves known to be used by wild females. The total number of clutches was similar for the two groups. The percent hatch for each clutch was then recorded.

The results of the study are presented as an interaction plot in Fig. 4-6. Indicated are the average percent hatches for clutches of each size at each location. The points indicate mean hatch for each clutch size, with the lines connecting points from the same geographic origin. The crossing of the lines indicates that opposite results were obtained for Iowa and Puerto Rico, with a higher proportion of eggs in large clutches hatching in Puerto Rico and the reverse true in Iowa (differences significant at $P < 0.1$ and 0.01 respectively). The field evidence thus suggests that local differences in selection produce adaptive differences in clutch size. Preliminary observations suggest that predation pressure is more intense in Iowa and that scattered smaller clutches are somewhat less likely to be found by predators and parasites. Physical factors such as the washing away of eggs on the periphery of the clutch may be more important in Puerto Rico.

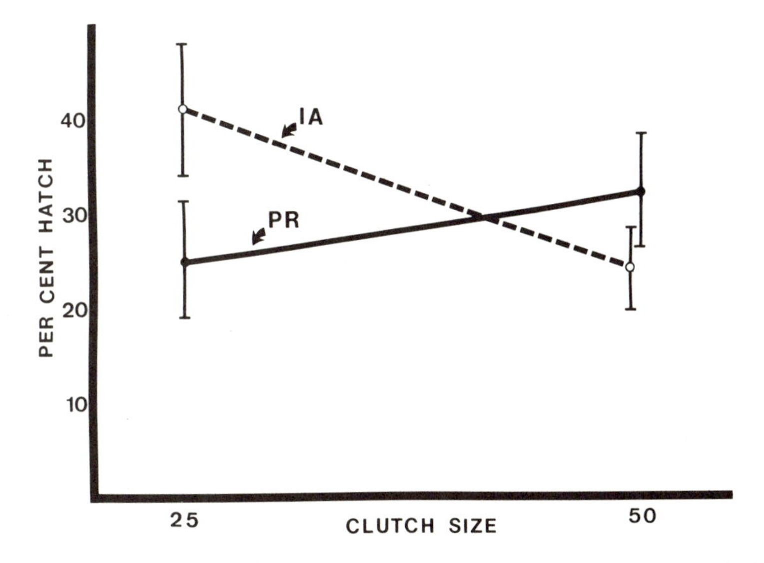

Figure 4-6. Interaction plot indicating survivorship of clutches of eggs of different sizes placed out on plants in Iowa (IA) and Puerto Rico (PR). Crossing of the lines indicates opposite effects at the two sites: i.e., more eggs hatch from larger clutches in Puerto Rico and from smaller ones in Iowa.

6 Discussion

The evolution of a migratory strategy has clear advantages for *O. fasciatus* populations in temperate continental North America. Migration in the spring allows exploitation of a superabundant but only seasonally available milkweed crop. In the autumn, migration in combination with diapause permits escape to more favorable climates. Since the bugs cannot overwinter, an absolute selective advantage accrues to any individuals that escape to breed elsewhere eventually (Dingle et al. 1977). We do not know at present what proportion of the individuals may succeed in doing so, but the chances must be increased considerably by the potential to enter reproductive diapuse for long periods. Diapause in short days is characteristic of Iowa bugs at a temperature of 23 °C (Fig. 4-4), and tethered flight durations indicate migratory capability (Table 4-2).

The generally equable climate of Puerto Rico permits continuous flowering and fruiting of the primary host plant. These conditions obviate the necessity for a migration-diapause strategy in this *O. fasciatus* population, and both are essentially absent. The reduced flight (Table 4-2) is consistent with a model which predicts gradual loss of dispersal genotypes as these leave and are not replaced by immigration (Roff 1975). As dispersers decline, their descendants further decrease relative to nondispersers, while dispersing offspring are reduced by mating with nondispersers until eventually a stable low frequency of dispersers is reached. The generality of the model with respect to insect populations is discussed in greater detail in Dingle (1978, see also Dingle et al. 1980b) and Järvinen and Vepsäläinen (1976). Sufficient flight capability is retained in the Puerto Rico population to allow local movement to exploit asynchronously fruiting patches of milkweed. Small scale "migrations" are thus important components of *O. fasciatus* life histories on this large tropical island. Similar patterns of flight and ecology were found for other species of tropical *Oncopeltus* (Blakley 1977, Dingle et al. 1980b).

In contrast to both Iowa and Puerto Rico, which have predictable climatic regimes, albeit for different reasons, Florida presents *O. fasciatus* with an unpredictable environment. Variation is among years and among sites within years. It is not surprising that, with respect to diapause, the responses of these insects are highly variable. The bug with a considerable capacity to delay reproduction may have a selective advantage one year, while one with early reproduction may have the advantage the next. The variance in diapause expected is evident in Fig. 4-5. Extensive discussions with similar reasoning for predicting high phenotypic (and genotypic) variances in the face of environmental uncertainty are presented by Dingle et al. (1977), Hoffman (1978) and Istock (1978, Chapter 7). The necessity at some times for flight between relatively distant breeding sites could maintain the observed intermediate levels of flight capacity. Occasional exchange with populations in continental North America to the north could also be a factor. Unfortunately the relative contributions of these two potential influences on flight are virtually impossible to measure.

The differing behavioral responses of the three *O. fasciatus* populations are reflected in their life table statistics. Under conditions inducing diapause-free development in both the Iowa and Puerto Rico samples, there is little difference in these statistics, especially as they contribute to population growth parameters such as r and R_0

(Table 4-4). Where a difference in these latter is observed, it is in the Florida bugs. In this case the chief contributing factor is the induction of diapause in a portion of the sample, resulting in delayed reproduction and reduction in total fecundity. The reduced egg output per female is a consequence of the increased probability of dying, by chance if nothing else, before completing maximum possible egg production. Iowa bugs which do enter diapause also experience considerable reproductive delay, with consequent effects in population growth (Dingle 1968). The "decision" by an individual to diapause or begin reproducing immediately requires trade-offs, which are obviously more readily assessable the more predictable and reliably cued the environmental milieu. The critical role played by behavior in life history scheduling should be apparent.

There is a difference between Iowa and Puerto Rico bugs in pattern of egg laying, clutches of 25 eggs every day for Iowa and clutches of 50 for Puerto Rico every other day. The difference is not reflected in r or R_0, which are essentially the same. In the various extensive theoretical discussions of the evolution of life histories (reviewed by Stearns 1976), some form of r or the "Malthusian parameter" is usually taken as a measure of fitness, and the phenotype is expressed as this fitness. Specific life history characters which may be undergoing selection, often in conflicting directions, are not usually considered (Livdahl 1979). The problems with this kind of reasoning are apparent in the clutch sizes and interclutch intervals noted here. These traits are under differing selective pressures at the different geographic locations (Fig. 4-6), and clearly contribute to fitness in that they influence offspring survival. Yet they do not, in this instance, contribute to the Malthusian measure of fitness, although, to the extent that they influence offspring survival, they could influence variance in r in natural populations.

The implications of variations in generation time caused by behavioral responses such as diapause and migration for life histories have been largely overlooked. They are now, however, beginning to receive attention (Nichols et al. 1976, Livdahl 1979, Taylor 1980). For example, Livdahl's (1979) model for the selection of life cycle delays predicts longer generation times with environmental uncertainty; a reduction of r is likely under these conditions. Our empirical data from *O. fasciatus* provide instances where reduction of r is of adaptive advantage (Table 4-4; Dingle 1968). The maintenance of high proportions of additive genetic variance by varying selection on diapause and migration in *Oncopeltus* and other insects (Dingle et al. 1977, Istock 1978) also has important implications for the development of life history theory.

Acknowledgments. Beth Alden, David Baldwin, Nigel Blakley, Bill Blau, and Ed and Elizabeth Klausner have all assisted with various aspects of this study. Bill Blau provided the data on contrasts in egg survival between Iowa and Puerto Rico and portions of the Puerto Rico census data. My research on *Oncopeltus* life histories is supported by the U.S. National Science Foundation (DEB-77-26412).

References

Blakley, N. R.: Evolutionary responses to environmental heterogeneity in milkweed bugs (*Oncopeltus,* Hemiptera, Lygaeidae). Ph.D. Thesis, University of Iowa, 1977.

Caldwell, R. L.: A comparison of the dispersal strategies of two milkweed bugs, *Oncopeltus fasciatus* and *Lygaeus kalmii.* Ph.D. Thesis, University of Iowa, 1969.

Caldwell, R. L.: A comparison of the migratory strategies of two milkweed bugs. In: Experimental Analysis of Insect Behaviour. Barton Browne, L. (ed.). New York: Springer-Verlag, 1974.

Derr, J. A., Alden, B. M., Dingle, H.: Insect life histories in relation to migration, body size and host plant array: a comparative study of *Dysdercus.* J. Anim. Ecol. (1981) (in press).

Dingle, H.: The relation between age and flight activity in the milkweed bug, *Oncopeltus.* J. Exp. Biol. 42, 269-283 (1965).

Dingle, H.: Life history and population consequences of density, photoperiod, and temperature in a migrant insect, the milkweed bug *Oncopeltus.* Am. Nat. 102, 149-163 (1968).

Dingle, H.: Migration strategies of insects. Science 175, 1327-1335 (1972).

Dingle, H.: The experimental analysis of migration and life history strategies in insects. In: Experimental Analysis of Insect Behaviour. Barton Browne, L. (ed.). New York: Springer-Verlag, 1974a.

Dingle, H.: Diapause in a migrant insect, the milkweed bug *Oncopeltus fasciatus* (Dallas) (Hemiptera: Lygaeidae). Oecologia 17, 1-10 (1974b).

Dingle, H.: Migration and diapause in tropical, temperate, and island milkweed bugs. In: Evolution of Insect Migration and Diapause. Dingle, H. (ed.). New York: Springer-Verlag, 1978.

Dingle, H.: Adaptive variation in the evolution of insect migration. In: Movement of Highly Mobile Insects: Concepts and Methodology in Research. Rabb, R. L., Kennedy, G. G. (eds.). Raleigh, N.C.: North Carolina State Univ. Press, 1979.

Dingle, H., Brown, C. K., Hegmann, J. P.: The nature of genetic variance influencing photoperiodic diapause in a migrant insect, *Oncopeltus fasciatus.* Am. Nat. 111, 1047-1059 (1977).

Dingle, H., Alden, B. A., Blakley, N. R., Kopec, D., Miller, E. R.: Variance in photoperiodic response within and among species of milkweed bugs (*Oncopeltus*). Evolution 34, 356-370 (1980a).

Dingle, H., Blakley, N. R., Miller, E. R.: Variation in body size and flight performance in milkweed bugs (*Oncopeltus*). Evolution 34, 371-385 (1980b).

Dixon, A. F. G., Wratten, S. D.: Laboratory studies on aggregation, size and fecundity in the black bean aphid, *Aphis fabae.* Scop. Bull. Entomol. Res. 61, 97-112 (1971).

Hoffman, R. J.: Environmental uncertainty and evolution of physiological adaptation in *Colias* butterflies. Am. Nat. 112, 999-1015 (1978).

Istock, C. A.: Fitness variation in a natural population. In: Evolution of Insect Migration and Diapause. Dingle, H. (ed.). New York: Springer-Verlag, 1978.

Järvinen, O., Vepsäläinen, K.: Wing dimorphism as an adaptive strategy in water striders (*Gerris*). Hereditas 84, 61-68 (1976).

Kennedy, J. S.: A turning point in the study of insect migration. Nature (London) 189, 785-791 (1961).

Kennedy, J. S.: Insect dispersal. In: Insects, Science, and Society. Pimentel, D. (ed.). New York: Academic Press, 1975.

Klausner, E.: Environmental variability and the ecology of the milkweed bug in South Florida. M.Sc. Thesis, Univesity of Iowa, 1979.

Klausner, E., Miller, E. R., Dingle, H.: *Nerium oleander* as an alternative host plant for South Florida milkweed bugs, *Oncopeltus fasciatus*. Ecol. Entomol. 5, 137-142 (1980).

Landahl, J. T., Root, R. B.: Differences in the life tables of tropical and temperate milkweed bugs, genus *Oncopeltus* (Hemiptera: Lygaeidae). Ecology 50, 734-737 (1969).

Laughlin, R.: Capacity for increase; a useful population statistic. J. Anim. Ecol. 34, 77-91 (1965).

Lewontin, R. C.: Selection for colonizing ability. In: Genetics of Colonizing Species. Baker, H. G., Stebbins, G. L. (eds.). New York: Academic Press, 1965.

Livdahl, T. P.: Environmental uncertainty and selection for life-cycle delays in opportunistic species. Am. Nat. 113, 835-842 (1979).

MacArthur, R. H., Wilson, E. O.: The Theory of Island Biogeography. Princeton, N.J.: Princeton Univ. Press, 1967.

Nichols, J. D., Conley, W., Batt, B., Tipton, A. R.: Temporally dynamic reproductive strategies and the concept of r- and K-selection. Am. Nat. 110, 995-1005 (1976).

O'Rourke, F. A.: Hybridization and systematics of Western Hemisphere species of milkweed bugs of the genus *Oncopeltus* (Hemiptera: Lygaeidae). Ph.D. Thesis, University of Connecticut, Storrs, 1977.

Rabb, R. L., Kennedy, G. G. (eds.): Movement of Highly Mobile Insects: Concepts and Methodology in Research. Raleigh, N.C.: North Carolina State Univ. Press, 1979.

Ralph, C. P.: Effect of host plant density on populations of a specialized, seed-sucking bug *Oncopeltus fasciatus*. Ecology 58, 799-809 (1977).

Roff, D. A.: Population stability and the evolution of dispersal in a heterogeneous environment. Oecologia 19, 217-237 (1975).

Root, R. B., Chaplin, S. J.: The life-styles of tropical milkweed bugs, *Oncopeltus* (Hemiptera: Lygaeidae) utilizing the same hosts. Ecology 57, 132-140 (1976).

Solbreck, C.: Migration, diapause, and direct development as alternative life histories in a seed bug, *Neacoryphus bicrucis*. In: Evolution of Insect Migration and Diapause. Dingle, H. (ed.). New York: Springer-Verlag, 1978.

Stearns, S. C.: Life history tactics: A review of the ideas. Q. Rev. Biol. 51, 3-47 (1976).

Taylor, F. Optimal switching to diapause in relation to the onset of winter. Theor. Pop. Biol. 18, 125-133 (1980).

Taylor, L. R., Taylor, R. A. J.: Aggregation, migration, and population mechanics. Nature (London) 265, 415-421 (1977).

Taylor, L. R., Taylor, R. A. J.: The dynamics of spatial behaviour. In: Population Control by Social Behaviour. Ebling, R. J., Stoddart, D. M. (eds.). London: Institute of Biology, 1978.

Vepsäläinen, K.: Wing dimorphism and diapause in Gerris: Determination and adaptive significance. In: Evolution of Insect Migration and Diapause. Dingle, H. (ed.). New York: Springer-Verlag, 1978.

Ziegler, J. R.: Evolution of the migration response: Emigration by *Tribolium* and the influence of age. Evolution 30, 579-592 (1976).

Chapter 5

Latitudinal Variation in the Life Histories of Insects Occupying Disturbed Habitats: A Case Study

WILLIAM S. BLAU

1 Introduction

Most of our ideas about how temperate and tropical regions differ have been derived from comparisons of climax communities. Such comparisons have led to generalizations regarding latitudinal variation in community stability, the intensity of biological interactions, the degree of species specialization, and the direction of life history evolution (e.g., Dobzhansky 1950, Klopfer and MacArthur 1960, Cody 1966, MacArthur and Wilson 1967). We are becoming increasingly aware that disturbance is an important and increasingly prevalent process in tropical as well as temperate environments (Connell 1978, Hubbell 1979, Garwood et al. 1979, Veblen 1979). In order to obtain a complete picture of how ecological and evolutionary pressures vary with latitude, we need to consider the community organization of organisms that exploit disturbed habitats, and the extent to which such communities contribute to the total biological assemblage. It is uncertain, a priori whether an understanding of disturbed-habitat communities will support the earlier conclusions drawn from climax communities or provide new perspectives.

We may begin to understand the biological implications of environmental disturbance by evaluating the nature of the disturbing process itself. Over what spatial scale does it occur, with what frequency or predictability, and at what intensity? Landslides, treefalls, severe storms and agriculture, for example, differ from one another both quantitatively and qualitatively in their impact on biological communities. The degree and rate of community recovery will also vary.

Localized disturbance inevitably leads to patchiness in habitat. Given sufficient information about the rates at which patches are created, the distribution of new patch sizes, their dispersion in space, and rates of recovery, a picture of the habitat structure of the environment as it relates to a particular colonizing species may be obtained (Levin and Paine 1974). If the dimensions that define habitat favorability can be recognized, it is important to understand how they vary within and among patches. A theoretical framework exists from which predictions may then be drawn and tested regarding individual species' strategies for exploiting a habitat (Southwood 1962, 1977,

Dingle 1972, Giesel 1976, Stearns 1976, Taylor and Taylor 1977, Solbreck 1978, Vepsäläinen 1978, Whittaker and Goodman 1979). The interactions between evolved life history parameters and extrinsic environmental conditions determine the population characteristics of a species and its position with respect to other members of the community.

In this paper, I summarize the results of comparative studies on the black swallowtail butterfly, *Papilio polyxenes* Fabr., in temperate and tropical environments (New York and Costa Rica) (Blau 1978, 1980). This species occurs in early successional habitats from southern Canada to northern South America. The studies have focused primarily on the ecological and evolutionary consequences of utilizing disturbed habitats. The results lead to a discussion of current ideas about the evolution of body size in insects, and I employ those ideas to generate hypotheses concerning latitudinal trends in the size of heliothermic species. This work provides a new perspective on the nature of differences between temperate and tropical environments—one that should be incorporated into our thinking about geographic variation in species and life history patterns.

2 The Study Organism

The life histories of both New York and Costa Rican populations of *P. polyxenes* follow the same basic scheme. Larvae may develop on any of a variety of host plant species belonging to the family Umbelliferae. About 3-5 weeks are required for egg maturation and growth of larvae through the final (fifth) instar. Adults eclose about 2 weeks later, unless pupal development is delayed by diapause. Females mate shortly after eclosion, and then range widely in open habitats, laying their eggs singly on appropriate hosts. In the laboratory, peak egg production occurs 2-5 days after eclosion, with 34-87 ovipositions per day (range of individual female averages over the 4-day period, N = 36). Reproductive output declines steadily over the next 3 or 4 weeks, by which time most females have died. Neither adult survival nor the temporal pattern of reproduction differs greatly between populations (Blau 1978).

3 Environments

3.1 Climate and Dispersion of Habitat

The tropical population discussed here occurred near the town of Turrialba, on the Atlantic slope of the Cordillera Central of Costa Rica. The temperate zone population was studied near the town of Brooktondale in central New York State. In both locations, *P. polyxenes* exploits habitats that are discontinuous in time and space.

Near Turrialba, the mean monthly temperature is nearly constant throughout the year while rainfall is seasonal (Fig. 5-1). The natural climax vegetation is premontane wet forest, at places in transition to tropical moist forest (Tosi 1969). At present most of the land is either cultivated for coffee and sugar cane or is used for pasture. *P. polyxenes* occurs in localized colonies associated with the only available host plant, *Spananthe paniculata* Jacq. This is an early successional forb, which germinates about 2

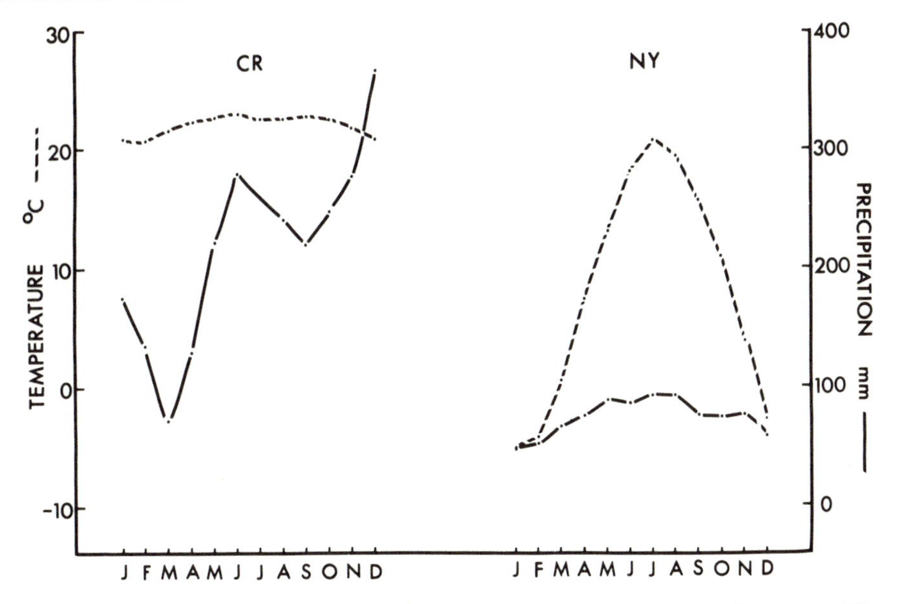

Figure 5-1. Mean monthly temperature and precipitation near Turrialba, Costa Rica, and Brooktondale, New York. Data are from the Central Agronómico Tropical de Investigación y Enseñanza (CR) and the Division of Atmospheric Sciences, Department of Agronomy, Cornell University (NY).

weeks after any disturbance that eliminates existing vegetation or provides a new substrate for germination (e.g., plowing, fire or riparian mud deposits). Black swallowtail females begin ovipositing on the plants when the first flowers are produced, about 6 weeks after germination. The butterflies continue to utilize the plants for up to 4 months before the latter senesce. Two or three distinct generations can be completed during this time period. They appear to be synchronized by the date at which the host plants first become acceptable to *P. polyxenes* females. Conditions suitable for the growth of *S. paniculata* may occur locally during any time of the year, including the relatively short dry season. Favorable habitat for *P. polyxenes* is therefore available at all times, although its distribution in space may be somewhat unpredictable, and breeding occurs throughout the year.

Patches of host plant in New York are more synchronous (Fig. 5-2). The seasonal patterns of temperature and rainfall for this area contrast sharply with those for Costa Rica (Fig. 5-1) and limit the annual growing season to a period from late May through the end of September. The climax vegetation is a mosaic of temperate deciduous beech-maple and oak-hickory forest. Only about one half of the land is currently maintained for agriculture, the remainder being in various stages of secondary succession. Among the several umbelliferous host species available, wild carrot, *Daucus carota* L., is by far the most widespread and commonly utilized. This biennial is most important during the first 5 years of secondary succession, and achieves its greatest density and coverage in the second year, when it may be dominant (Bard 1952). It is frequently found, however, in waste areas, roadsides, or in other sites where disturbance is recurrent, and so may persist locally for 20 years or more. As a result, the spatial distribution of habitat for *P. polyxenes* is more predictable in New York than in Costa Rica, while the precise timing and duration of the growing season fluctuates from year to year.

In New York, *P. polyxenes* overwinters in pupal diapause. Adults eclose sometime during May or early June. Two generations (broods) of larvae usually follow. Most pupae produced at the end of the second brood enter diapause, but in some years a partial third brood is observed. Colonies of *P. polyxenes* in New York are not as highly localized or well-defined as those in Costa Rica. Males defend mating territories at sites chosen on the basis of topographic features rather than host plant presence (Lederhouse 1978). Females are more vagile, wandering among feeding and oviposition sites. Even where host plants abound, densities of eggs and larvae are usually very low.

3.2 Habitat Favorability

Figure 5-2 indicates that the degree of habitat favorability varies both among patches and within patches over time. Differences among patches, besides those due to age, are likely to result from variability in microenvironments, surrounding habitats, the nature of the disturbances that create a patch and, in Costa Rica, the time of year. Differences within a patch over time are reflected in the survivorship curves of eggs and larvae from successive generations at the same site (Fig. 5-3). Two points should be noted: (1) egg and larval survival is generally lower in New York than in Costa Rica (Fig. 5-3, center), and (2) in both locations the juvenile survivorship decreases from one generation to the next.

Total mortality in these experiments was partitioned into component rates of mortality due to small predators, large predators, and drowning or dislodging by rain. The survivorship of free individuals subject to all sources of mortality was compared with those for individuals enclosed in one of two types of cages. Cages made of chicken wire (0.5-inch mesh) excluded large predators, and identical ones with sheets of plastic at the top eliminated the effects of both large predators and rain. In both populations, survivorships decreased over time because rates of predation increased (Fig. 5-4). In New York the increase was attributable to large predators alone, while in Costa Rica

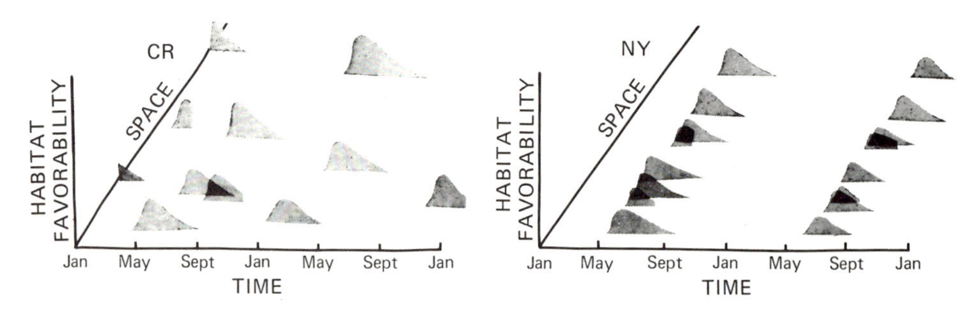

Figure 5-2. A conceptualization of the spatiotemporal pattern of habitat favorability for *P. polyxenes* in Costa Rica and New York. Shaded portions represent host plant patches; patch size is not indicated. Truncated curves that are not at the edge of the figure represent patches destroyed by a repeated disturbance. See text for discussion.

both large and small predators were responsible. In some instances, locally high densities of juveniles and adult males also contributed to the decline in patch favorability in Costa Rica (Blau 1980).

Pupal survival differs considerably between the two locations. In Costa Rica, about 85% of all pupae followed in the field survived to produce adults. Evidence of parasitism was rare. Data from this study and those of previous years indicate that pupal survival in New York is highly variable and often low (< 50%) due to high rates of parasitism.

4 Life Histories

4.1 Observations and Predictions

Because habitat patches are relatively short-lived [Southwood's (1977) H/τ = 2-3, i.e., a patch can support only two or three generations] and decrease in quality over time, *P. polyxenes* must escape from old patches and locate new ones on a regular basis. Differences in the regularity and reversibility of changes associated with patchiness in each environment have resulted in the evolution of different mechanisms for escape (Southwood 1962, 1977). Since favorable habitat is available throughout the year in Costa Rica, the irregular and destructive changes that result from localized disturbances favor escape in space by dispersal. Such escape is not as effective a means of increasing

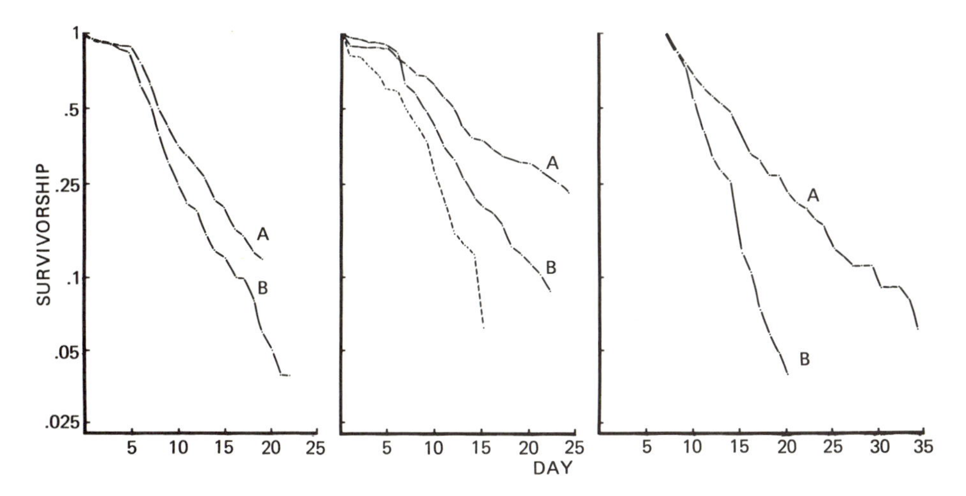

Figure 5-3. Survivorship (ℓ_x) curves for eggs and larvae, derived from records for approximately 30 individuals per day. Oviposition occurs at day 0 and eggs hatch at about day 6 or 7. Curves labeled A and B represent the first and second generations, respectively, at a given site. Left: Site 1 (Montaña), Costa Rica, September and October, 1976. Center: Site 2 (Bosque), Costa Rica, February and March, 1976. Curve for New York Brood II, 1977, shown for comparison (broken line, terminated after third larval instar). Right: Broods I and II in New York, June and August, 1977. Data are presented for only the first through fourth (A) or third (B) larval instars.

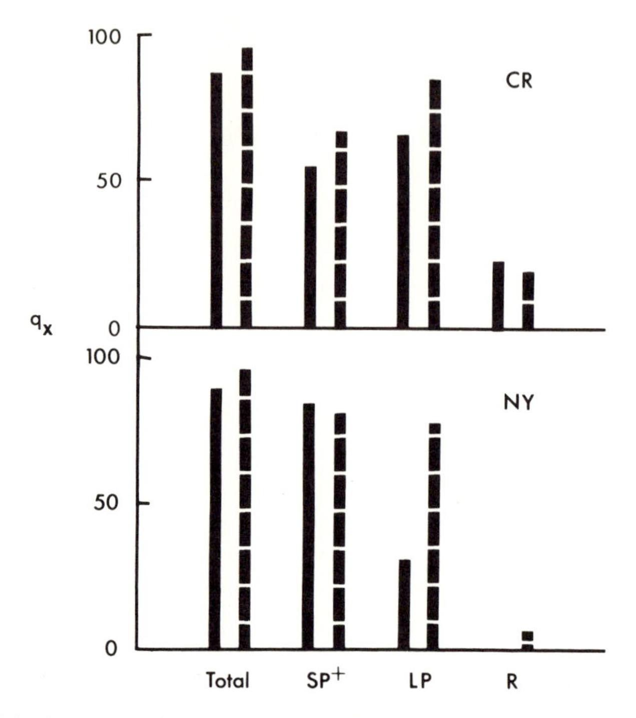

Figure 5-4. Absolute mortality rates, expressed as percentage mortality (q_x) due to small predators and miscellaneous causes (SP^+), large predators (LP) and rain (R). Data are for the first (solid bars) and second (broken bars) generations at Site 1 in Costa Rica (egg through final larval instar) and in New York (first through third larval instars). Absolute mortality rates are those that would have resulted from one particular source of mortality in the absence of all others.

reproductive success in New York, where the favorabilities of all patches are maximal in the spring and converge at zero by the end of the growing season. Although females may disperse in order to minimize the risk of total reproductive failure (den Boer 1968) or to capitalize on variability in quality among host plant patches, the regular and reversible habitat changes associated with seasonality favor escape by a facultative diapause in situ cued by short day-lengths (Oliver 1969, Blau 1978).

Ecological escape leads ultimately to the colonization of new habitats; therefore, we should expect to find in both populations adaptations conventionally attributed to colonizers—i.e., early age at first reproduction, high fecundity, and a concentration of that fecundity early in adult life (Cole 1954, Lewontin 1965, MacArthur and Wilson 1967). To the extent that the spatiotemporal patterns of habitat differ between New York and Costa Rica, however, the above life history variables ought to be tailored within each population to best effect colonization in its respective environment. In Costa Rica, those individuals that locate and colonize new habitats most rapidly on a continual basis will contribute most to future generations. For organisms with fecundity as high as that of *P. polyxenes* (several hundred eggs), a decrease in the age at first reproduction is the most effective means of further increasing the number of descendants produced over the long term and would, therefore, be the most likely adaptation to occur.

The same argument does not apply to the New York population. Because the breeding cycle is interrupted after two generations, an individual's reproductive success is not determined by the rate at which progeny would be produced under a regime of continual reproduction. It is rather determined by the relative numbers of progeny that successfully enter diapause at the end of the annual growing season (in most cases, the end of the second generation). Therefore, a reduction of the age at first reproduction would not necessarily be beneficial to New York individuals. Selection should favor other adaptations, such as an increase in mean fecundity, that would maximize reproductive output over the course of two generations. Because fecundity is positively correlated with adult size in many insects, including *P. polyxenes* (Blau 1978), the ages at pupation and at first reproduction may be delayed in order to provide a longer period for growth. [A large reduction in the age at first reproduction might be beneficial if it allowed the completion of an additional generation during the growing season. This becomes more likely as the length of the growing season increases, and should result in discontinuities in the expression of life history traits over latitudinal clines. Such discontinuities have been observed for crickets in the Japanese archipelago (Masaki 1978)].

To summarize, selection for colonizing ability should favor early age at first reproduction, high fecundity, and high daily fecundity (m_x) early in adult life in both populations. The actual values which these variables assume will represent a trade-off between the advantages (higher fecundity) and disadvantages (delayed reproduction) of prolonging the term of larval development. It is hypothesized that the balance of selective forces acting upon the Costa Rican population will be shifted in comparison to that in New York, so as to favor earlier age at first reproduction at the expense of some fecundity. On the basis of this selective balance hypothesis, Costa Rican females are predicted to pupate earlier and to produce smaller, less fecund adults than New York females. Also, because of the sensitivity of the intrinsic rate of increase (r, a population parameter) to shifts in the age of first reproduction, the Costa Rican population should exhibit a higher value of r, as calculated from laboratory life history data.

4.2 Evidence Bearing on the Predictions

In order to test the above predictions, I measured the developmental and reproductive characteristics of representatives from each population in the field when feasible, and in detail under controlled laboratory conditions. Between the two sets of observations, an accurate picture emerges of the inheritable and expressed life history attributes for each population. Using laboratory-derived ℓ_x and m_x values, r was found by trial and error substitution in the Euler equation to be 0.136 for the Costa Rican population and 0.129 for the New York population. This difference is small, but in the anticipated direction, even though the sample of New York females in this experiment had unusually high fecundities. It is, in fact, due to the shorter development period of Costa Rican larvae, which results in earlier maturation and decreases the age at first reproduction by 5 days (Fig. 5-5, top). The durations of the egg stage, the pupal stage, and the prereproductive adult period differ only slightly between populations, and the fecundity schedules are similar, with reproductive output peaking within a few days of adult maturity.

These results, although in harmony with the selective balance hypothesis, are misleading. The results of laboratory feeding experiments indicate that differences in de-

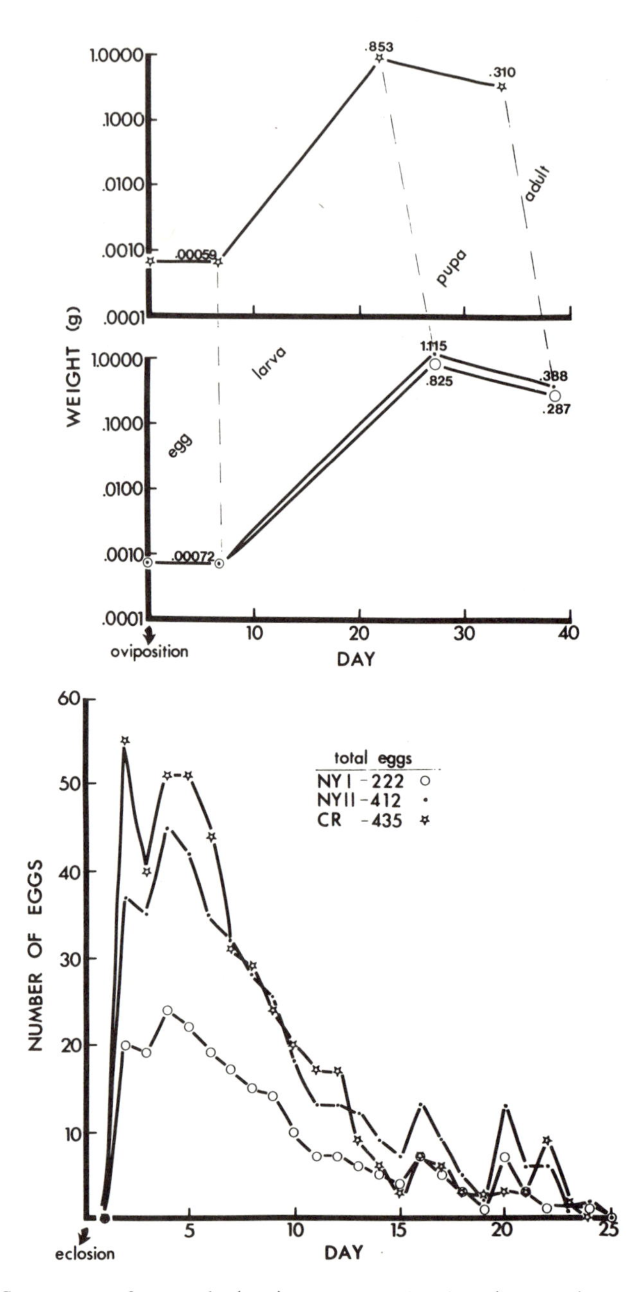

Figure 5-5. Summary of growth (top) and reproduction (bottom) in *P. polyxenes*. Symbols are defined in the figure. The weights of eggs, pupae and adults are given; note the logarithmic scale of the weight axis. The data are means or estimates based on laboratory and field observations.

velopment time are not the products of divergent evolution, but are imposed by differences in the effective nutrient content of the respective host plants (Fig. 5-6). Although the levels of energy, nitrogen and water in the leaf tissues of *S. paniculata* and *D. carota* are similar, all larvae are more efficient at extracting energy and nitrogen from and grow more rapidly on *S. paniculata* (Blau 1978). When reared on *D. carota,* larvae of both origins exhibit the same average rate of growth—14% per day by length.

If host plant effects are factored out so that growth rates are identical, it becomes evident that New York larvae are capable of producing adult females that are 25% larger than those from Costa Rica in the same amount of time. This is possible because New York larvae have an initial size advantage of nearly the same proportion: New York eggs are 22% larger than Costa Rican eggs. Observations indicate that this size difference is maintained after hatching and carried throughout larval development. This potential for larger adults to be produced in New York is realized only when larvae develop under photoperiods longer than 14.25 hours, as they do in the field during the spring generation. These larvae produce the adults of the summer generation which, although they are larger than Costa Rican adults, lay similar numbers of eggs because their eggs are larger (Fig. 5-5). Their offspring experience photoperiods of shorter than critical duration. The short-day photoperiodic cue effects a switch in the mode of development so that the resulting autumn pupae (1) undergo diapause, (2) are brown in color (West et al. 1972), and (3) are smaller than those of the previous generation and more similar in size to those from Costa Rica. The time required to complete development is not altered significantly; the late summer larvae apparently grow more slowly over the same period of time. In addition to this developmental switch, a direct physiological response to the warm temperatures during the summer larval generation may result in the production of smaller pupae (von Bertalanffy 1960, e.g., Miller 1977). As a consequence, spring adults are significantly smaller and their reproductive output is reduced considerably as compared with that of Costa Rican females or New York females later in the season. No photoperiodic response has been detected for the Costa Rican population.

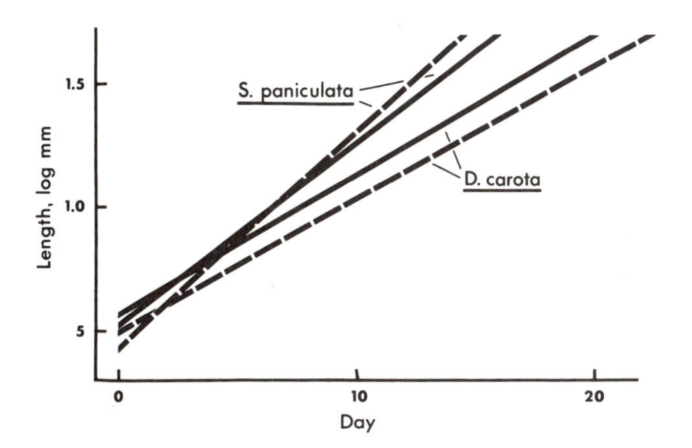

Figure 5-6. Larval growth, by length, on *S. panculata* and *D. carota* in the laboratory. Solid lines are for New York larvae and broken lines are for Costa Rican larvae. Day 0 is the day of hatching.

5 Interpretation of Life History Differences

5.1 General Considerations

The results of this life history analysis do not support the selective balance hypothesis. The later age at first reproduction in New York individuals is imposed by differences in host plant quality, and does not appear to be adaptive. The larger size of the summer-generation New York females does not confer a higher fecundity because of the parallel difference in egg size, and the smaller spring females actually lay fewer eggs than their similarly sized Costa Rican counterparts. Selection for colonizing ability does not appear to have led to differentiation in the reproductive patterns of the two populations. Either the underlying argument that led to the hypothesis is not justifiable, or other evolutionary pressures not taken into account have obscured the expected adaptive responses to each environment.

The most striking differences found between the New York and Costa Rican populations, apart from the presence or absence of a diapause response, are in egg size, body size, and the response of the latter to photoperiod. Size, being correlated with a variety of other life history traits, may itself be considered a key life history variable—an indicator of a species' relative position in the spectrum of possible reproductive strategies (Pianka 1970, Blueweiss et al. 1978, Western 1979). At the same time, it is subject to a complex assortment of other selective pressures and environmental influences that may bear little relation to strategies for reproduction, and may complicate life history analyses. The following discussion reviews the life history and other ecological consequences of variation in insect size. It leads to the development of an hypothesis for explaining seasonal and geographic size variation in *P. polyxenes.*

5.2 The Regulation of Body Size in Insects

The variety of factors demonstrated or hypothesized to influence the evolution of adult size in insects is summarized in Fig. 5-7. The ecological impact of body size is most often realized through its developmental, physiological or physical correlates. There is also one way, at least, in which size itself may directly influence an individual's fitness: by determining the number of predators for which that individual is a potential prey (Schoener and Janzen 1968, Wilson 1975).

Several of the physiological correlates of adult size involve the ability of the organism to maintain or regulate its internal milieu in the face of varying external stresses. Among mammals, body size is positively correlated with the length of time an individual can withstand starvation, and the same is likely to be true for insects. Blakley (1977) has found significant correlations between the survival of starved adults and their weight at eclosion within four species of milkweed bugs, genus *Oncopeltus.* Also, among organisms that are similar in shape, the ratio of body surface area:volume decreases as size increases. Since the relative proportion of surface area represents the available interface for physical exchanges between the organism and its environment, it has important consequences for the ability to regulate internal water and heat content. Schoener and Janzen (1968) found that insect samples from drier habitats con-

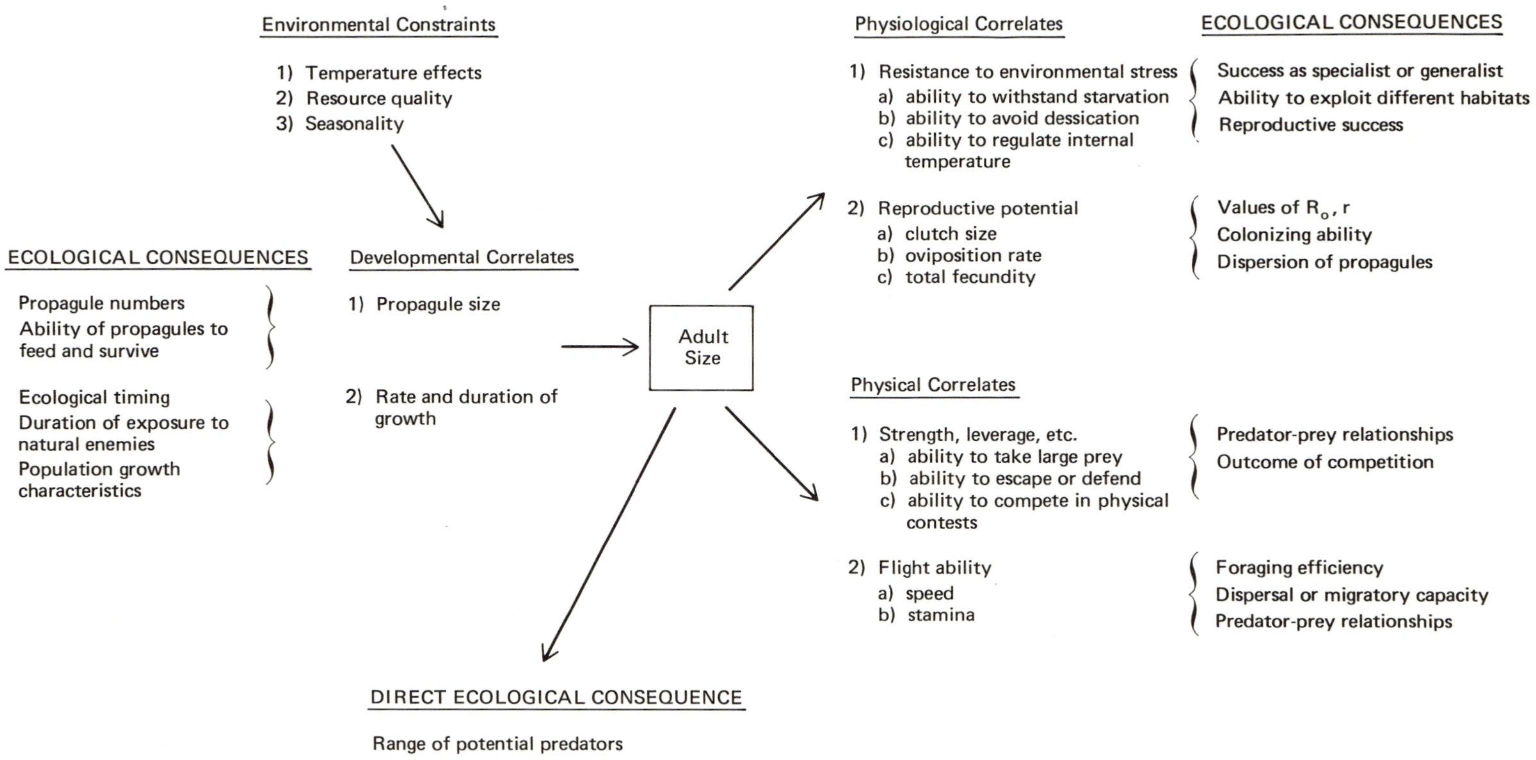

Environmental Constraints

1) Temperature effects
2) Resource quality
3) Seasonality

ECOLOGICAL CONSEQUENCES

Propagule numbers
Ability of propagules to feed and survive

Ecological timing
Duration of exposure to natural enemies
Population growth characteristics

Developmental Correlates

1) Propagule size

2) Rate and duration of growth

Adult Size

Physiological Correlates

1) Resistance to environmental stress
   a) ability to withstand starvation
   b) ability to avoid dessication
   c) ability to regulate internal temperature

2) Reproductive potential
   a) clutch size
   b) oviposition rate
   c) total fecundity

ECOLOGICAL CONSEQUENCES

Success as specialist or generalist
Ability to exploit different habitats
Reproductive success

Values of R_o, r
Colonizing ability
Dispersion of propagules

Physical Correlates

1) Strength, leverage, etc.
   a) ability to take large prey
   b) ability to escape or defend
   c) ability to compete in physical contests

2) Flight ability
   a) speed
   b) stamina

Predator-prey relationships
Outcome of competition

Foraging efficiency
Dispersal or migratory capacity
Predator-prey relationships

DIRECT ECOLOGICAL CONSEQUENCE

Range of potential predators

Figure 5-7. Some correlates of adult size in insects, and their ecological consequences.

tained larger species, as might be expected if larger insects are more resistant to desiccation. The influence of body size on rates of heat exchange and the ability of insects to thermoregulate is well-documented (Bartholomew and Heinrich 1973, Heinrich 1974, May 1976), and will be discussed in greater detail in the next section.

In general terms, Wasserman and Mitter (1978) proposed that large body size in feeding-generalist moths serves as a buffer against the broad range of environmental stresses they are thought to experience. The effects of size on water and thermal balance provide specific mechanisms whereby such individuals may indeed be buffered. Derr, Alden and Dingle (in prep.) present a different argument based on the relationship between body size and the ability to withstand starvation. They believe that large size is an adaptation of feeding specialists among bugs in the genus *Dysdercus,* enhancing migratory capability, explosive reproduction (see below), and the ability to survive diapause when food availability is low. Regardless of whether or not size is related to degree of feeding specialization, it is clear that it can influence the effectiveness with which an organism may exploit different habitats.

One of the more obvious ways in which size may be related to fitness is through its effect on reproductive potential. Total fecundity, oviposition rate and clutch size are often positively correlated with adult female size within insect species (Morris 1963, Englemann 1970, Blau 1978, Dunlap-Pianka 1979). Differences in the fecundity schedule have a direct effect on the net replacement rate or the intrinsic rate of increase, the latter parameter being widely employed as an index of fitness, especially with regard to colonizing ability (e.g., Lewontin 1965, Dingle 1974, Solbreck 1978). Variability in clutch size or oviposition rate may also play a role in determining the dispersion of offspring with respect to resources and natural enemies.

The physical correlates of adult size fall into two categories. The first involves attributes that are determined by an organism's strength and ability to manipulate. With particular reference to predators, the size of an individual determines to a large extent the size distribution of food items available to it, and may affect the rate at which resources can be gathered. Consequently, differentiation in predator body size can result from selection for optimal foraging ability or from competitive displacement with respect to food resources (Schoener 1969, Wilson 1975, Calow 1977, Van Zant et al. 1978). The strength advantage associated with larger size may also play a role in the ability of prey to defend themselves or escape from predators, as suggested by Schoener and Janzen (1968) and Calow (1977), and the outcome of intraspecific (e.g., Johnson 1981) or interspecific (e.g., Kikuchi 1965) competition where physical contests may actually occur.

Another physical correlate of adult size is flight ability. Recent studies have begun to elucidate the relationship between size and the speed and duration of insect flight. Pyke (1978) found that larger-sized bumblebees fly faster while foraging for nectar and hence take less time to fly between inflorescences than do smaller conspecifics. Because the energy cost of flight increases with size, the observed size of bumblebees is predicted to be some optimal value which maximizes the net rate of energy uptake. Dingle and Arora (1973) and Dingle et al. (1980) measured flight of tethered individuals for several species of bugs (genera *Dysdercus* and *Oncopeltus,* resp.), and found size to be correlated with both flight duration and the presumed importance of flight in the species' life history. It is unresolved whether the relationship between size and flight is causal, e.g., body size determines flight stamina, or is the result of more complex interactions among size, the ability to withstand starvation, reproductive capacity, and the ecological pressures promoting flight (Roff 1978, Dingle et al. 1980).

Because all growth in insects occurs in the preadult stages, adult size cannot vary independently of developmental characteristics such as propagule size and growth rate or duration; neither can developmental characteristics vary without influencing adult size. The body size of adults must be interpreted in light of the direct environmental constraints and ecological pressures imposed on the developmental stages. For example, ambient temperature and resource quality can influence adult size through their effects on larval growth rates (von Bertalanffy 1960, Nijhout and Williams 1974, Rausher 1979). The restriction of development time in annual species by the length of the growing season may cause size to vary with latitude (Schoener and Janzen 1968, Masaki 1978). Any ecological pressure that exerts an influence on propagule size also has the potential to influence adult size. If preadult growth is assumed to be exponential, the biomass of eggs or newly hatched larvae is converted to adult biomass by the application of a growth rate over the developmental period (excluding egg, pupal or inactive stages). If the rate and duration of growth are held constant, any change in initial biomass should be passed on to the adult stage in equal proportion (Blau 1978). In general, small propagule size should be favored so that the number of progeny produced per unit reproductive investment is maximized, to the point where further reduction in size is prohibited by other pressures relating to the ability of offspring to survive and gather food (Labine 1968, Dunlap-Pianka 1979). The logic of this argument requires that the rate or duration of growth varies to offset the differences in propagule size and produce equal-sized adults. But the duration of the growth period and the rate of growth, to the extent that it influences duration, are subject to their own ecological constraints. Larval growth must be timed not only to coincide with the availability of resources, but also to ensure that proper photoperiodic cues are received during a particular life stage, or that adults eclose when their own resources (including mates) are available (Tauber and Tauber 1978). If mortality rates for larvae are higher than those for pupal or adult stages, there may be a premium on completing larval development as rapidly as possible. Finally, development time will influence a variety of population traits including generation time, voltinism, age at first reproduction and the intrinsic rate of increase. The relationships among developmental parameters, adult size and fecundity are obviously complex, and have not been adequately dealt with by life history theorists.

5.3 The Case for Thermoregulation

On the basis of what is already known about the life history and habits of *P. polyxenes* it is possible to rule out a priori several factors mentioned above, for example, most of those dealing with predator-prey interactions. The importance of the reproductive correlates of size differences between New York and Costa Rican adults has already been ruled out because the parallel difference in egg size nullifies any effect on oviposition rate or fecundity. Any argument pertaining to larval duration also suffers because (a) interpopulation differences in larval duration are due primarily to differences in the host plants, (b) second-generation larvae in New York take no less time to complete development, even though they pupate at a smaller size. Although the importance of dispersal in the tropical population has been stressed, New York females are also highly mobile, and flight ability is likely to be just as important there as in Costa Rica for locating mates, for moving among feeding and oviposition sites, and for spreading the risk of reproductive failure by dispersing eggs in space. Therefore, it is not clear why selection for flight ability should favor size differentiation within or between populations.

In contrast, the effect adult size has on the ability of a heliotherm such as *P. polyxenes* (Rawlins 1980) to regulate its internal body temperature is likely to have direct and substantial consequences for fitness. The adult stage is highly specialized for rapid mating and reproduction, nearly all vegetative functions being relegated to the preadult stages. But vigorous flight activity can only occur within a specified range of physiologically acceptable internal temperatures. Thoracic temperatures between 28 and 32 °C are preferred by *P. polyxenes* males in New York, and butterflies in the field thermoregulate behaviorally within this range when ambient temperatures are between 14 and 22 °C (Rawlins 1980). Only gliding or soaring flight is possible as temperatures fall below 14 °C, and ultimately physiological processes are slowed to the point where sustained flight is impossible. At ambient temperatures above 22 °C, cooling behavior is inadequate to maintain thoracic temperature within the preferred range, and heat stress may occur. Consequently, the amount of time available to an individual for activities related to reproduction varies with that individual's ability to maintain its internal temperature within the 28-32 °C range. Those who are less able than others to achieve and maintain optimal temperatures, or who take longer to do so, will suffer a selective disadvantage in the form of reduced reproductive output (Clench 1966, Watt 1968, Shapiro 1976).

Several avenues of adaptation have been shown to be effective in altering thermal balance in insects. These include physiological (metabolic thermogenesis), behavioral (range of activities from sun-basking through shade-seeking), and morphological (melanism, hairiness, body size) adaptations. For example, Hoffman (1973) has found that seasonal polyphenism in the size and wing melanism of *Colias eurytheme* is under photoperiodic control. "Cold weather" forms produced under short photoperiods exhibit smaller size and increased pigmentation, allowing them to heat more rapidly. The influence of body size on thermal balance in insects has been discussed in greater detail by May (1976). The greater surface:volume ratio of smaller insects leads to higher rates of heating and cooling and lower equilibrium body temperatures. Larger-bodied insects can attain higher equilibrium temperatures that are more stable when environmental conditions fluctuate. Although their rates of heat exchange are lower, they have a potentially greater ability to regulate internal temperature.

Based on the above information, the fact that behavioral thermoregulation may detract from time available for other activities, and the observed inter- and intrapopulation differences in the size of *P. polyxenes,* I have constructed the following hypotheses:

(1) Where the environment is cool [with respect to the voluntary activity range of the species (Watt 1968)] and variable (with respect to day and night temperatures or the intensity of solar radiation during the day) (e.g., temperate spring), smaller size

is optimal because it allows individuals to heat up rapidly and make use of short intervals during which thermal conditions are favorable. The size adaptation may be coupled with increased melanism in order to raise the equilibrium body temperature. Large individuals will be at a disadvantage, as they will heat more slowly and may rarely achieve optimal internal temperatures.

(2) Where the environment is warm but variable (e.g., temperate summer), larger individuals will more readily achieve optimal body temperatures, and will have the advantage of being better able to regulate their temperature to correct for either overheating or overcooling. Once normal activity levels have been reached, larger individuals are more buffered against short-term environmental changes that may restrict the activity of smaller individuals.

(3) Where the environment is warm and constant (e.g., lowland tropics), overheating will be the primary stress, especially in open habitats where individuals are fully exposed to the sun. Smaller size is again optimal because the equilibrium body temperature will be less above ambient, and because convective cooling during flight in smaller individuals will be more rapid and effective (Digby 1955). Larger individuals will be less able to dissipate heat generated by the muscles during flight, and will cool more slowly at rest. The size adaptation may be accompanied by decreased pigmentation.

Implicit in these hypotheses is the assumption that it is adult size which is regulated by selection and not that of some other developmental stage to which adult size may be allometrically scaled in time. Thus, the larger size of eggs in New York may be a secondary adaptation for producing larger adults without prolonging larval development.

Is the observed size variation in *P. polyxenes* of sufficient magnitude to have a significant effect on thermal balance? The data presented in Table 5-1 demonstrate that it is. Several investigators have related, both theoretically and empirically, the thermoregulatory properties of an insect to the diameter of its thorax (see below). Thoracic measurements were obtained for several pinned specimens of *P. polyxenes* from New York Brood II and Costa Rica. There was virtually no overlap in size between the two samples, and the 1.4-mm mean difference is highly significant (Mann-Whitney U-test,

Table 5-1. Thoracic diameters of pinned specimens of *P. polyxenes* and estimates of the thermal properties of individuals from New York (Brood II) and Costa Rica

	New York Brood II	Costa Rica
Thoracic diameter in mm (N)	4.6 (5)	3.2 (6)
Rate of temperature increase (°C/sec)[a]	0.3	0.5-0.6
Equilibrium temperature excess (above ambient) (°C)		
50 cm/sec wind[b]	11.0	8.2
300 cm/sec wind[c]	3.0	1.8

[a] After a 6 °C rise from ambient temperature, estimates based on data of Papageorgis (1975) for butterflies exposed to direct sunlight.
[b] Estimates based on data of Church (1960) for denuded insects under artificial radiation.
[c] Estimates based on data of Digby (1955) for Diptera and Hymenoptera under artificial radiation.

$P < 0.005$). The impact of a difference of this magnitude is evident from an examination of published data on size and thermal conductance. Butterflies the size of Costa Rican black swallowtails heat up roughly twice as fast as those the size of New York swallowtails (cf. Papageorgis 1975) (Table 5-1). Equilibrium temperatures will vary with morphology and ambient conditions, but can differ by at least as much as 3 °C over the observed size range (cf. Digby 1955, Church 1960). A sample of thoracic diameters from New York Brood I females was too small to compare statistically with one from Brood II females (N = 2), but the mean size difference (3.9 vs. 4.8 mm) was similar to that between Costa Rican and New York Brood II individuals.

Further evidence from other species suggests that the thermoregulation hypotheses may have general applicability. Seasonal variability in melanism is widespread among the diurnal Lepidoptera, with spring and autumn forms generally being darker than summer forms (Shapiro 1976). Shapiro describes an experiment in which he reared different seasonal forms of *Pieris protodice* and released them at different times of the year. Rates at which the two forms were recaptured suggest that each is more successful than the other during its appropriate season. It is unknown whether this result is due to differences in pigmentation or in size between the two forms since, as in *C. eurytheme,* the darker form is also smaller (Scudder 1889). The same is true for at least three other seasonally polyphenic species: *Pieris rapae, P. napi* and *Polygonia interrogationis.* Watt's (1969) data for *C. eurytheme* also indicate that within a seasonal form males warm up more rapidly than females, who are larger (Hoffman 1973). With respect to latitudinal trends, Heinrich (1972) found that *Precis villida,* a denizen of open habitats near the equator, operates with thoracic temperatures of 41-42 °C. This is near the upper limit of tolerance, and cooling is probably effected by forced convection during flight in this relatively small species (cf. *Papilio aegeus,* same study). May (1976) has also discussed the association of small size with tropical dragonflies that nearly conform to ambient temperatures. Downes (1964, 1965) notes that individuals from more northern populations of arctic Lepidoptera are generally smaller than those from populations in more favorable areas, although he suggests that this is the result of differences in resource availability.

In conclusion, the argument for thermoregulatory ability as a determinant of body size, although in need of further testing, is compelling. Wide-ranging heliothermic species are subject to thermal stresses that vary over space and time; they must adapt to those stresses or suffer physiological inefficiency and reduced viability. Estimates of the heritability of body size in at least one insect (*O. fasciatus,* Dingle et al. 1980) indicate that moderate amounts of additive genetic variance exist for this trait. Thus, body size has the potential for evolving rapidly in response to changing selective pressures, in contrast to the complex reorganization of temperature-sensitive physiological processes that would probably be required for an evolutionary change in the preferred range of temperatures for activity. If new evidence fails to provide a basis for rejecting the hypotheses presented here, they may prove to be of general use in understanding size patterns in a variety of heliothermic species.

6 Concluding Remarks

For insects, latitudinal patterns in the ecology and evolution of disturbed-habitat species may be expected to differ in a great variety of ways from those of species inhabiting more permanent habitats. In general, the idea that disturbances resulting from seasonality in temperate environments impose "r-selection" is misleading, since breeding is discontinuous and stable age distributions are rarely achieved (Taylor 1979). Combinations of life history attributes that result in a maximal value of r may be more characteristic of species exploiting disturbed habitats in aseasonal tropical environments. There, a continual supply of temporary refuges from natural enemies and locally overexploited habitats can provide an effectively unlimited supply of resources in the context of an uninterrupted breeding season. The catch, of course, is that new habitat patches must be found and colonized, and this implies high mobility within a widely dispersed population—a characteristic not traditionally associated with tropical species.

Temperate populations may also undergo cycles of colonization on an annual basis, but need not experience selection for life history attributes that confer a maximal value of r. Although predictions based on this argument have not held for the populations of *P. polyxenes* described here, they may do so for populations from other portions of this species' range, or for other organisms whose reproductive patterns are not obscured by size adaptations to unknown or unrelated pressures (e.g., perhaps crickets, Masaki 1978). Overall, patterns in the reproductive adaptations of temperate species to seasonal environments are likely to be the products of complex interactions among the length of the growing season, environmental changes that occur during the growing season, and the evolutionary history of the organism. In addition to latitudinal variation in the presence or timing of diapause, patterns of more finely tuned adjustments to seasonality should occur, especially seasonal polyphenism in size, color, and perhaps reproductive output.

Blueweiss et al. (1978) and Western (1979) have discussed the general importance of body size as a morphological correlate of an array of life history traits and population parameters. I have argued above that latitudinal variation in insect body size may reflect a shifting balance in the combinations of life history attributes that result in the greatest contribution of progeny to future generations. At the same time, it is apparent that selective pressures that are related only indirectly to life history strategies can also vary seasonally or geographically to produce variation in body size. I believe this to be the case for the black swallowtail butterfly, where selection for optimal thermal balance may explain observed size differences between spring and summer adults in New York, and between adults from New York and Costa Rica. Detailed investigations of several other populations throughout the range of this species will be required in order to determine whether or not selection for life history tactics also plays a role in producing latitudinal variation in body size.

Acknowledgments. I am grateful to the following people for lending support of one kind or another to this research: May Berenbaum, Lorraine Contardo, Paul Feeny, Robert Lederhouse, Mark Rausher, Richard Root, Joseph and Yvonne Saunders and

Kristi Wharton. Paul Feeny provided the data on thoracic widths in *P. polyxenes.* Patricia Blau, Robert Denno, Hugh Dingle, Edward Klausner, John Rawlins and an anonymous reviewer suggested improvements in the manscript. I am especially grateful to Patricia Blau for assisting in all aspects of this work. The research in Costa Rica was made possible by the cooperation of the Centro Agronómico Tropical de Investigación y Enseñanza. Financial support was provided by NSF Grants BMS 74-09868, BMS 75-15282, and DEB 76-20114 to Paul Feeny.

This research was conducted while I was a graduate student at Cornell University, and the original manuscript was prepared while I was a postdoctoral trainee in the Neurobehavioral Sciences Program at the University of Iowa, Iowa City.

References

Bard, G. E.: Secondary succession on the piedmont of New Jersey. Ecol. Monogr. 22, 195-215 (1952).

Bartholomew, G. A., Heinrich, B.: A field study of flight temperatures in moths in relation to body weight and wing loadings. J. Exp. Biol. 58, 123-135 (1973).

Blakley, N.: Evolutionary responses to environmental heterogeneity in milkweed bugs (*Oncopeltus:* Hemiptera: Lygaeidae). Ph.D. Thesis, University of Iowa, 1977.

Blau, W. S.: A comparative study of the ecology, life histories, and resource utilization of temperate and tropical populations of the black swallowtail butterfly, *Papilio polyxenes* Fabr. Ph.D. Thesis, Cornell University, 1978.

Blau, W. S.: The effect of environmental disturbance on a tropical butterfly population. Ecology 61, 1005-1012 (1980).

Blueweiss, L., Fox, H., Kudzma, V., Nakashima, D., Peters, R., Sams, S.: Relationships between body size and some life history parameters. Oecologia 37, 257-272 (1978).

Calow, P.: Ecology, evolution and energetics: a study in metabolic adaptation. Adv. Ecol. Res. 10, 1-62 (1977).

Church, N. S.: Heat loss and the body temperature of flying insects. II. Heat conduction within the body and its loss by radiation and convection. J. Exp. Biol. 37, 186-212 (1960).

Clench, H. K.: Behavioral thermoregulation in butterflies. Ecology 47, 1021-1034 (1966).

Cody, M. L.: A general theory of clutch size. Evolution 20, 174-184 (1966).

Cole, L. C.: The population consequences of life history phenomena. Q. Rev. Biol. 29, 103-137 (1954).

Connell, J. H.: Diversity in tropical rain forests and coral reefs. Science 199, 1302-1310 (1978).

den Boer, P. J.: Spreading of risk and stabilization of animal numbers. Acta Biotheor. 18, 165-194 (1968).

Digby, P. S. B.: Factors affecting the temperature excess of insects in sunshine. J. Exp. Biol. 32, 279-298 (1955).

Dingle, H.: Migration strategies of insects. Science 175, 1327-1335 (1972).

Dingle, H.: The experimental analysis of migration and life-history strategies in insects. In: Experimental Analysis of Insect Behaviour. Barton Browne, L. (ed.). New York: Springer-Verlag, 1974, pp. 329-342.

Dingle, H., Arora, G.: Experimental studies of migration in bugs of the genus *Dysdercus.* Oecologia 12, 119-140 (1973).

Dingle, H., Blakley, N. R., Miller, E. R.: Variation in body size and flight performance in milkweed bugs (*Oncopeltus*). Evolution 34, 356-370 (1980).

Dobzhansky, T. H.: Evolution in the tropics. Am. Sci. 38, 209-221 (1950).

Downes, J. A.: Arctic insects and their environment. Can. Entomol. 96, 279-307 (1964).

Downes, J. A.: Adaptations of insects in the arctic. Ann. Rev. Entomol. 10, 257-274 (1965).

Dunlap-Pianka, H. L.: Ovarian dynamics in *Heliconius* butterflies: correlations among daily oviposition rates, egg weights, and quantitative aspects of oogenesis. J. Insect Physiol. 25, 741-749 (1979).

Englemann, F.: The Physiology of Insect Reproduction. New York: Pergamon Press, 1970.

Garwood, N. C., Janos, D. P., Brokaw, N.: Earthquake-caused landslides: a major disturbance to tropical forests. Science 205, 997-999 (1979).

Giesel, J. T.: Reproductive strategies as adaptations to life in temporally heterogeneous environments. Ann. Rev. Ecol. Syst. 7, 57-79 (1976).

Heinrich, B.: Thoracic temperatures of butterflies in the field near the equator. Comp. Biochem. Physiol. 43A, 459-467 (1972).

Heinrich, B.: Thermoregulation in endothermic insects. Science 185, 747-756 (1974).

Hoffman, R. J.: Environmental control of seasonal variation in the butterfly *Colias eurytheme*. I. Adaptive aspects of a photoperiodic response. Evolution 27, 387-397 (1973).

Hubbell, S. P.: Tree dispersion, abundance, and diversity in a tropical dry forest. Science 203, 1299-1309 (1979).

Johnson, L. K.: Sexual selection in a brentid weevil. Evolution (1981) (in press).

Kikuchi, T.: Role of interspecific dominance-subordination relationship on the appearance of flower-visiting insects. Sci. Rep. Tohuku Univ. (Biol.) 31, 275-296 (1965).

Klopfer, P. H., MacArthur, R. H.: Niche size and faunal diversity. Am. Nat. 94, 293-300 (1960).

Labine, P. A.: The population biology of the butterfly, *Euphydras editha*. VIII. Oviposition and its relation to patterns of oviposition in other butterflies. Evolution 22, 799-805 (1968).

Lederhouse, R. C.: Territorial behavior and reproductive ecology of the black swallowtail butterfly, *Papilio polyxenes asterius* Stoll. Ph.D. Thesis, Cornell University, 1978.

Levin, S. A., Paine, R. T.: Disturbance, patch formation, and community structure. Proc. Nat. Acad. Sci. USA 71, 2744-2747 (1974).

Lewontin, R. C.: Selection for colonizing ability. In: The Genetics of Colonizing Species. Baker, H. G., Stebbins, G. L. (eds.). New York: Academic Press, 1965, pp. 79-94.

MacArthur, R. H., Wilson, E. O.: The Theory of Island Biogeography. Princeton, N.J.: Princeton Univ. Press, 1967.

Masaki, S.: Seasonal and latitudinal adaptations in the life cycles of crickets. In: Evolution of Insect Migration and Diapause. Dingle, H. (ed.). New York: Springer-Verlag, 1978, pp. 72-100.

May, M. L.: Thermoregulation and adaptation to temperature in dragonflies (Odonta: Anisoptera). Ecol. Monogr. 46, 1-32 (1976).

Miller, W. E.: Weights of *Polia grandis* pupae reared at two constant temperatures (Lepidoptera: Noctuidae). Great Lakes Entomol. 10, 47-49 (1977).

Morris, R. F.: The dynamics of epidemic spruce budworm populations. Memoirs Entomol. Soc. Am. 31, 1-332 (1963).

Nijhout, H. F., Wllliams, C. M.: Control of moulting and metamorphosis in the tobacco hornworm, *Manduca sexta* (L.): growth of the last-instar larva and the decision to pupate. J. Exp. Biol. 61, 481-491 (1974).

Oliver, C. G.: Experiments on the diapause dynamics of *Papilio polyxenes.* J. Insect Physiol. 15, 1579-1589 (1969).

Papageorgis, C.: Mimicry in neotropical butterflies. Am. Sci. 63, 522-532 (1975).

Pianka, E. R.: On r- and K-selection. Am. Nat. 104, 592-597 (1970).

Pyke, G. H.: Optimal body size in bumblebees. Oecologia 34, 255-266 (1978).

Rausher, M. D.: Larval habitat suitability and oviposition preference in three related butterflies. Ecology 60, 503-511 (1979).

Rawlins, J. E.: Thermoregulation by the black swallowtail butterfly, *Papilio polyxenes* (Lepidoptera: Papilionidae). Ecology 61, 345-357 (1980).

Roff, D. A.: Size and survival in a stochastic environment. Oecologia 36, 163-172 (1978).

Schoener, T. W.: Models of optimal size for solitary predators. Am. Nat. 103, 277-313 (1969).

Schoener, T. W., Janzen, D. H.: Notes on environmental determinants of tropical versus temperate insect size patterns. Am. Nat. 102, 207-224 (1968).

Scudder, S. H.: The Butterflies of the Eastern United States and Canada with Special Reference to New England. III Vols. Cambridge, Mass.: W. H. Wheeler, 1889.

Shapiro, A. M.: Seasonal polyphenism. Evol. Biol. 9, 259-333 (1976).

Solbreck, C.: Migration, diapause, and direct development as alternative life histories in a seed bug, *Neacoryphus bicrucis.* In: Evolution of Insect Migration and Diapause. Dingle, H. (ed.). New York: Springer-Verlag, 1978, pp. 195-217.

Southwood, T. R. E.: Migration of terrestrial arthropods in relation to habitat. Biol. Rev. 37, 171-214 (1962).

Southwood, T. R. E.: Habitat, the templet for ecological strategies? J. Anim. Ecol. 46, 337-365 (1977).

Stearns, S. C.: Life history tactics: A review of the ideas. Q. Rev. Biol. 51, 3-47 (1976).

Tauber, M. J., Tauber, C. A.: Evolution of phenological strategies in insects: a comparative approach with eco-physiological and genetic considerations. In: Evolution of Insect Migration and Diapause. Dingle, H. (ed.). New York: Springer-Verlag, 1978, pp. 53-71.

Taylor, F.: Convergence to the stable age distribution in populations of insects. Am. Nat. 113, 511-530 (1979).

Taylor, L. R., Taylor, R. A. J.: Aggregation, migration, and population mechanics. Nature (London) 265, 415-421 (1977).

Tosi, J. A., Jr.: Mapa ecologico, Republica de Costa Rica. San José, Costa Rica: Centro Cientifico Tropical, 1969.

Van Zant, T., Poulson, T. L., Kane, T. C.: Body-size differences in carabid cave beetles. Am. Nat. 112, 229-234 (1978).

Veblen, T. T.: Structure and dynamics of *Nothofagus* forests near timberline in south-central Chile. Ecology 60, 937-945 (1979).

Vepsäläinen, K.: Wing dimorphism and diapause in *Gerris*: determination and adaptive significance. In: Evolution of Insect Migration and Diapause. Dingle, H. (ed.). New York: Springer-Verlag, 1978, pp. 218-253.

von Bertalanffy, L.: Principles and theory of growth. In: Fundamental Aspects of Normal and Malignant Growth. Nowinski, W. W. (ed.). Holland: Elsevier, 1960, pp. 137-252.

Wasserman, S. S., Mitter, C.: The relationship of body size to breadth of diet in some Lepidoptera. Ecol. Entomol. 3, 155-160 (1978).

Watt, W. B.: Adaptive significance of pigment polymorphisms in *Colias* butterflies. I. Variation of melanin pigment in relation to thermoregulation. Evolution 22, 437-458 (1968).

Watt, W. B.: Adaptive significance of pigment polymorphisms in *Colias* butterflies. II. Thermoregulation and photoperiodically controlled melanin variation in *Colias eurytheme*. Proc. Nat. Acad. Sci. USA 63, 767-774 (1969).

West, D. A., Snellings, W. M., Herbek, T. A.: Pupal color dimorphism and its environmental control in *Papilio polyxenes asterius* Stoll (Lepidoptera: Papilionidae). N.Y. Entomol. Soc. 80, 205-211 (1972).

Western, D.: Size, life history and ecology in mammals. Afr. J. Ecol. 17, 185-204 (1979).

Whittaker, R. H., Goodman, D.: Classifying species according to their demographic strategy. I. Population fluctuations and environmental heterogeneity. Am. Nat. 113, 185-200 (1979).

Wilson, D. S.: The adequacy of body size as a niche difference. Am. Nat. 109, 769-784 (1975).

Chapter 6

Geographic Variation of the Diapause Response in the European Corn Borer

WILLIAM B. SHOWERS

1 Introduction

Ostrinia nubilalis (Hubner) is an introduced insect species belonging to the order Lepidoptera that has had a significant impact on the production of *Zea mays* L., field maize, and particularly sweet maize, as well as other host plants that possess stems (North Cent. Reg. Comm. NC-105, 1972). Therefore, studies of geographic variation in this introduced species may be of great practical as well as basic significance.

O. nubilalis (European corn borer, ECB) came to North America nearly 68 years ago, possibly in broom corn imported from central Europe (Caffrey and Worthley 1927). The species is holometabolous: egg, larva, pupa and adult (moth). The female moth usually deposits a cream-colored mass of eggs (ca. 25 eggs) on the underside of the leaf of a host plant. The eggs hatch in 4 days and the neonates immediately feed on the various parts of the plant. The insects progress through 5 larval molts and, usually after the third molt, bore into the stalk or stem of the host plant, where the larvae feed and grow to a length of approximately 3 cm. Larval feeding destroys both the phloem and xylem of the host plant and characteristically produces large cavities the length of the stalk or stem. Presently, ECB occur only in temperate regions. Therefore, the species enters diapause during late summer or early autumn and overwinters as fifth-instar larvae (last larval instar) in the stalks or stems of the host plants (Caffrey and Worthley 1927).

From the beginning, *O. nubilalis* populations in North America comprised at least two voltine types. As early as 1914, Vinal (1917) determined the presence of a bivoltine form of this species in Massachusetts, and shortly thereafter Felt (1919) reported a univoltine form near Lake Erie and in western New York. Although the bivoltine population of ECB normally occurs in Iowa (Harris and Brindley 1942), Showers and Reed (1971) found that warm temperatures during September and October would allow third-generation larvae to mature to the fifth instar and then diapause and survive the Iowa winter.

However, the effect of temperature on voltinism of this insect was first hypothesized by Arbuthnot (1949). Using historical mean annual temperature and normal annual precipitation data from 279 stations in the United States and 73 stations in Europe and Asia, he attempted to predict the geographical locations in the United States where one, two, three or more generations per year of ECB would occur (Fig. 6-1a). The expected generations compare favorably with the present distribution of yearly generations of ECB in North America (Fig. 6-1b). Arbuthnot obviously was correct

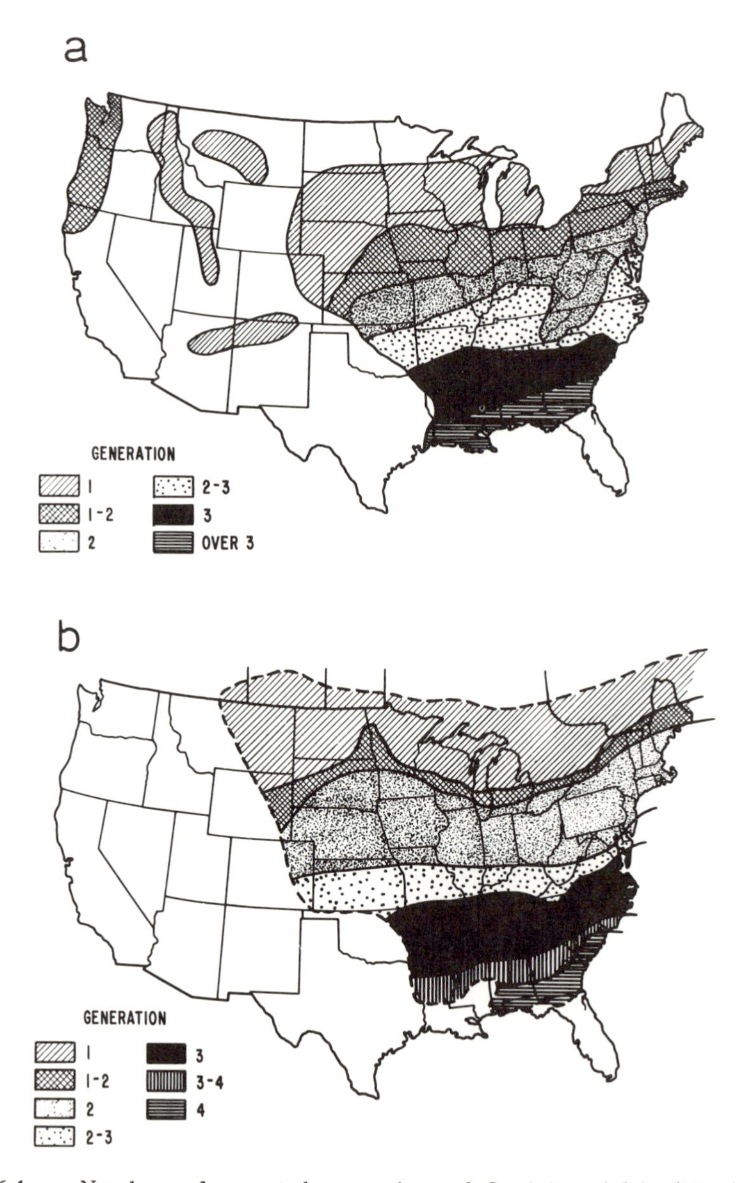

Figure 6-1. a: Numbers of expected generations of *Ostrinia nubilalis* (Hbn.) in the United States, based on climatic zones (Arbuthnot 1949). b: Approximate distribution of present generation zones of *Ostrinia nubilalis* (Hbn.) in the United States and Canada (Showers 1979).

that temperature is important in determining the number of yearly generations of ECB, but he was unaware of the importance of photoperiod. It was not until the late 1950s that Mutchmor and Beckel (1959) demonstrated that long photophase (Beck in 1962 determined that it was actually short scotophase) caused pupation (nondiapause), and that low temperature accompanied by short photophase caused diapause.

These earlier researchers opened the way for many studies on diapause and geographic variation of this pest of agricultural commodities. More recent studies have implicated juvenile hormone and ecdysiotropin as possible elements of the diapause system (Chippendale 1977), and have implied that induction or termination of diapause might fit within a hourglass framework (Skopic and Bowen 1976). A recent study has also shown that inheritance of a diapause trait follows a pattern substantially consistent with that of a sex-linked system (Reed et al. 1981).

The basic significance of the results of the studies presented in this chapter is that one or more environmental factors can affect geographic populations of a species differently, and that, once the effect is elucidated, the population response can be genetically manipulated. Thus, by increasing our knowledge of the causes of natural variation in populations, we hope to genetically engineer a serious agricultural pest insect species for management purposes.

2 Ecotypes

During the past 18 years several researchers have addressed themselves to the effect of diapause response on geographic variation among populations of ECB. Sparks et al. (1966a) studied ECB populations from Minnesota, Iowa and Missouri (north to south), and found that hybrids resulting from crosses in which at least one parent was of Minnesota origin showed an increased incidence of diapause in the laboratory and in the field. They therefore concluded that diapause in ECB is influenced by a polygenic response to photoperiod and temperature. Chiang et al. (1968) concluded that the southern population of ECB (Missouri) was more responsive to photoperiod and temperature than was the northern population of ECB (Minnesota).

Showers et al. (1972) expanded the research begun by Sparks et al. (1966a), and conducted field studies with the progeny of reciprocal matings of populations of ECB from the southern fringe, east coast, and northern fringe of the species' distribution. They found that, under natural photophases of 14 hr 22 min to 14 hr 41 min, Minnesota (univoltine) parentage has little influence on the reciprocal F_1 populations. However, delaying the infestation of field plots with ca. 60 neonate larvae per maize plant for 10 days allowed the first adult emergence to occur under natural photophases of 13 hr 55 min to 14 hr 12 min. Consequently, similar to the results of Sparks et al. (1966a), the incidence of diapause in F_1 populations was heavily influenced by the Minnesota (univoltine) parent (Fig. 6-2).

Of particular significance, however, was the effect of the Alabama male on the diapause response of the progeny of the reciprocal matings with Maryland and Minnesota females (1969 data in Fig. 6-2). Showers et al. (1972) suggested that the pupation response (indicating development to adulthood) as well as the diapause response (indicating inhibition of development) of ECB is sex-linked to the parental male. Later,

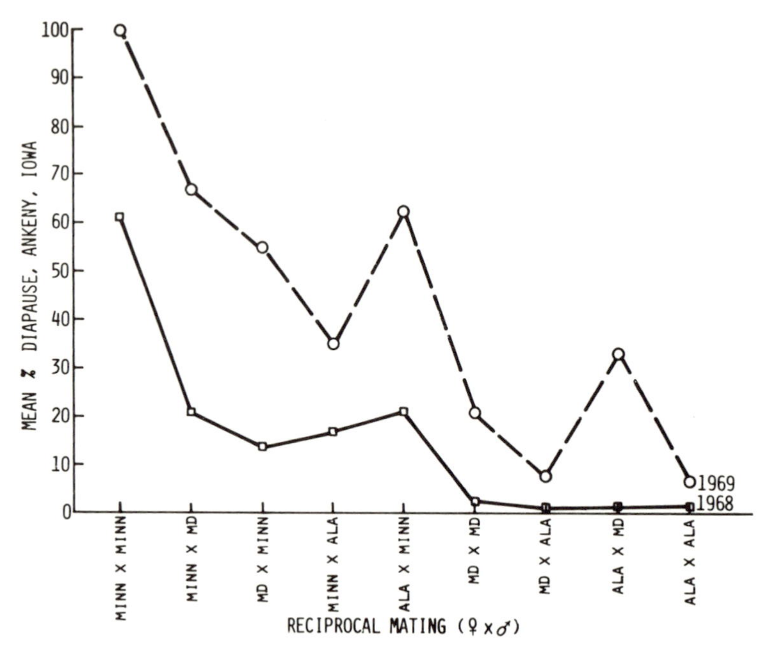

Figure 6-2. Percentage diapause in progeny of 9 matings of *Ostrinia nubilalis* (Hbn.), Ankeny, Iowa, 1968-69. Photophase at average first adult emergence was 25 min shorter in 1969 (Showers et al. 1972).

Reed et al. (1981) crossed a low-diapausing laboratory population and a low-diapausing field population to a naturally occurring, high-diapausing (univoltine) population of ECB. When the male parent was univoltine, the progeny had a higher frequency of diapause, a higher proportion of females in the diapause populations, and a lower proportion of females in the low-diapause populations than when the male parent was of a low-diapause population. Equally interesting was the fact that all progeny tended to respond, i.e., go into diapause or continue development through to adulthood, in the same way as the male parent. This suggestion of paternal influence would argue against cytoplasmic inheritance of the diapause trait, as cytoplasmic inheritance normally shows maternal influence. Reed et al. (1981) found that reciprocal differences were common. Therefore, it is not likely that autosomal inheritance controls the diapause trait. Reciprocal effects and paternal influence in ECB where the female is the heterogametic sex (Srb et al. 1965) suggest sex-linked inheritance.

Reed et al. (1981) calculated expected diapause frequencies, sex ratios of diapausing progeny, and sex ratios of nondiapausing progeny for sex-linked inheritance by a single nondominant gene. Diapause was designated d+ and nondiapause, d. Females were designated d+o (diapause) and do (nondiapause). Males were designated d+d+ (diapause), d+d (heterozygous), and dd (nondiapause). All d+d+ and d+o insects were assumed to diapause, all dd and do insects were assumed to pupate and develop into adults; likewise, 50 % of the d+d insects were assumed to diapause and 50% were assumed to pupate and develop into adults.

Once Reed et al. (1981) had calculated the expectations, the crosses were grouped into 6 phenotypes (Table 6-1). The expectations are compared with the actual results in Fig. 6-3. The discrepancies in the comparisons were explained by Reed et al. (1981) as follows: First, the removal of diapausing females, due to larval mortality, produced lower female:male diapause sex ratios than expected. This could account for the discrepancies in diapause sex ratios of phenotypes A, B, C and D. Second, the nondiapause parents (ND and GA) both produced a significant number of diapause progeny, indicating that these populations were not homozygous for the nondiapause gene. Thus, unexpected diapausing progeny occurred in phenotype F with a diapause sex ratio of 48:52 (none were expected). These diapausing progeny were probably a measure of the frequency of diapausing alleles in the nondiapause parents. Progeny from crosses of "nondiapause parents" with diapause alleles thus added both females and males to all crosses, though more evidently in phenotypes E and F. Similarly, the diapause parent (Q) produced some nondiapause progeny, though Q appeared to be nearly homozygous. However, this small frequency of nondiapause alleles in the diapause parent did produce some unexpected nondiapause insects (no expectation for nondiapause sex ratio, phenotype A in Fig. 6-3). The nondiapause alleles in the diapause parent could likewise explain the discrepancy between expected and observed for the nondiapause sex ratios of phenotype B, since it could cause some insects (both female and male) to pupate when they were expected to diapause. Finally, the nondiapause sex ratios of phenotypes A and B reflect relatively small numbers of insects, an indication that frequency of nondiapause alleles was lower in the diapause parent than were frequencies of diapause alleles in the nondiapause parents. Although Reed et al. (1981) indicate that neither the diapause parent nor the nondiapause parent were homozygous for the diapause response or the pupation response, there was a close relationship between results and expectations.

The NC-105, North Central Technical Committee, a group of scientists involved in interregional research on the ecology of ECB, then constructed a hypothesis, based on the results of Sparks et al. (1966a, 1966b), Chiang et al. (1968), Showers et al. (1972) and Reed et al. (1981), that the ECB in North America consists of univoltine, bivoltine and multivoltine populations. To test this hypothesis, Showers et al. (1975) conducted field studies at 7 locations from Minnesota to Georgia. The results show that diapause induction of ECB is directly related to the interaction between photoperiod (short day, long night) and cool temperature, regardless of the geographic source of the bivoltine or multivoltine population.

On the basis of the diapause response of ECB, Showers et al. (1975) have partitioned the ECB populations in North America into three ecotypes (Table 6-2). The diapause response of the three ecotypes, as determined experimentally, is presented in Fig. 6-4. The northern ecotype (represented by univoltine populations from Quebec and Minnesota) will go into summer diapause (August) at any latitude in the United States. The central ecotype (represented by populations from Iowa, Nebraska and Ohio) has a relatively low percentage summer diapause (40%) north of latitude 41°N. However, on or near 41°N, about 85% will diapause, and at 36°N the number of individuals in summer diapause again drops to about 45%. Further south (31°27'N), about 82% of the individuals of the central ecotype will go into summer diapause. The populations of ECB composing the southern ecotype (Alabama, Georgia, Missouri) sustained higher

Table 6-1. Calculated expectations for diapause response governed by a single, nondominant sex-linked gene (Reed et al. 1981)

Phenotype[a]	Crosses with identical expectations (female X male)	Expectations		
		% Diapause	Female:male nondiapause	Female:male diapause
A	QXQ, (NDXQ) (Q), (GAXQ) (Q)	100	–	50:50
B	NDXGA, GAXQ, (QXND) (Q), (QXGA) (Q)	75	0:100	67:33
C	(NDXQ) (NDXQ), (GAXQ) (GAXQ), (Q) (NDXQ) (Q) (GAXQ), (Q) (QXND), (Q) (QXGA)	62.5	67:33	40:60
D	(QXND) (QXND), (QXGA) (QXGA), (ND) (NDXQ) (GA) (GAXQ), (ND) (QXND), (GA) (QXGA)	37.5	40:60	67:33
E	QXND, QXGA, (NDXGA) (ND), (GAXQ) (GA)	25	67:33	0:100
F	NDXND, GAXGA, (QXND) (ND), (QXGA) (GA)	0	50:50	

[a] A = Quebec x Quebec, (Lab x Quebec) (Quebec), (Georgia x Quebec) (Quebec).
B = Lab x Georgia, Georgia x Quebec, (Quebec x Lab) (Quebec), (Quebec x Georgia) (Quebec).
C = (Lab x Quebec) (Lab x Quebec), (Georgia x Quebec) (Georgia x Quebec), (Quebec) (Lab x Quebec), (Quebec) (Georgia x Quebec), (Quebec) (Quebec x Lab), (Quebec x Georgia).
D = (Quebec x Lab) (Quebec x Lab), (Quebec x Georgia) (Quebec x Georgia), (Lab) (Lab x Quebec), (Georgia) (Georgia x Quebec), (Lab) (Quebec x Lab), (Georgia) (Quebec x Georgia).
E = Quebec x Lab, Quebec x Georgia, (Lab x Georgia) (Lab), (Georgia x Quebec) (Georgia).
F = Lab x Lab, Georgia x Georgia, (Quebec x Lab) (Lab), (Quebec x Georgia) (Georgia).

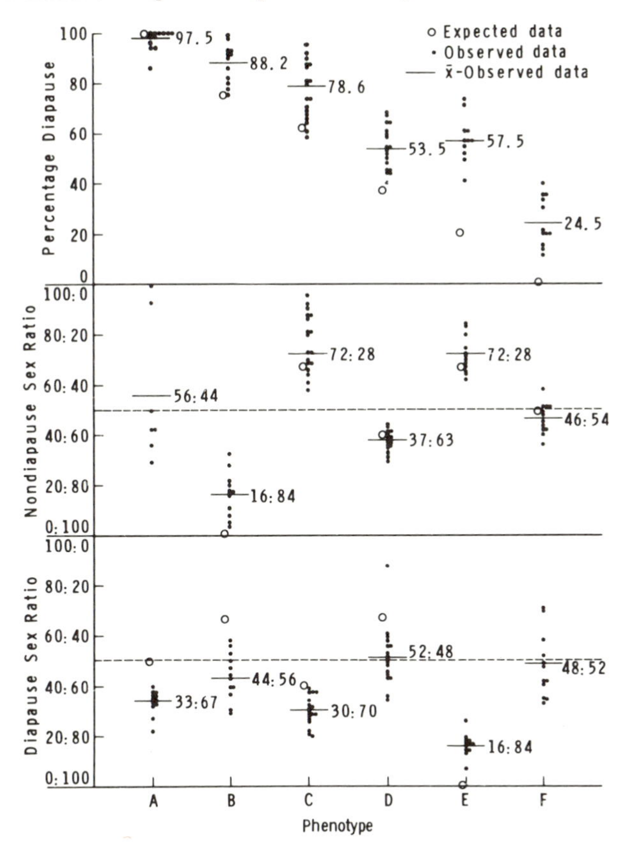

Figure 6-3. Expectations for sex-linked inheritance of the diapause trait in *Ostrinia nubilalis* (Hbn.) and observed results for larval development in 12-, 13:20- and 14:40-hr photophases. Expectations for single nondominant sex-linked gene; results expressed as the mean response of the progeny of each cross in each photophase (Reed et al. 1981).

percentage summer diapause near 41°N than they did north of this latitude, lower percentage diapause at 38°N and higher percentage diapause again near 36°N. However, further south at 31°N, diapause dropped to about 15%.

I interpret these data to mean that the short-day adapted, heat-sensitive, multivoltine populations (southern ecotype) of ECB avoid summer diapause along the Gulf Coast, but that the more northern bivoltine populations (central ecotype) are not able to efficiently use the abundant degree units of the lower South. Therefore, these populations (Iowa, Nebraska, Ohio) of ECB are severely affected by the short daylength (long night) and are forced into summer diapause. Near latitude 36°N, however, the warm temperatures of the region interact with the relatively short daylengths, so that the daylength is not critical. Thus, the central ecotype avoids summer diapause. At or near latitude 41°N the temperatures are relatively cool and, even though the daylengths are relatively long, cool temperatures allow the daylength to become critical and force the central ecotype into summer diapause. North of latitude 41°N, however, the very

Table 6-2. Ecotypes of *Ostrinia nubilalis* (Hbn.) in North America

Northern (univoltine)
Quebec
Minnesota
Wisconsin
Central (2 generations/yr)
Iowa
Maryland (mountains)
Nebraska
Ohio
Ontario
Southern (3 or more generations/yr)
Alabama
Georgia
Maryland (eastern shore)
Missouri

long days (short nights) compensate for the relatively cool temperatures and the photoperiod-sensitive central ecotype again avoids summer diapause.

How the interaction between photoperiod and temperature affects the diapause response of ECB is presented in Fig. 6-5. If a specific bivoltine or multivoltine population of ECB occurs north of the latitude where the interaction between photoperiod and temperature exerts the greatest influence on diapause for that population, the diapause response during summer would be low and few would diapause (north portion of Fig. 6-5). If a specific bivoltine or multivoltine population of ECB is on the latitude of greatest photoperiod-temperature interaction (nights relatively long, temperatures relatively cool), most of the population would diapause (center of Fig. 6-5). However,

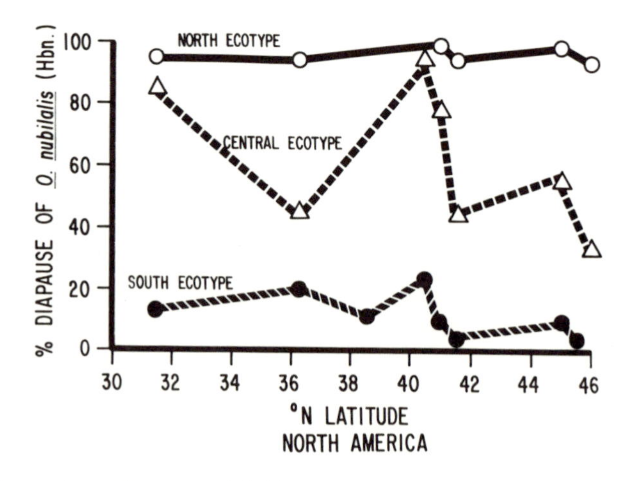

Figure 6-4. Percentage diapause of field-tested populations representing 3 ecotypes of *Ostrinia nubilalis* (Hbn.) at several latitudes in the United States. (Adapted from Showers et al. 1975.)

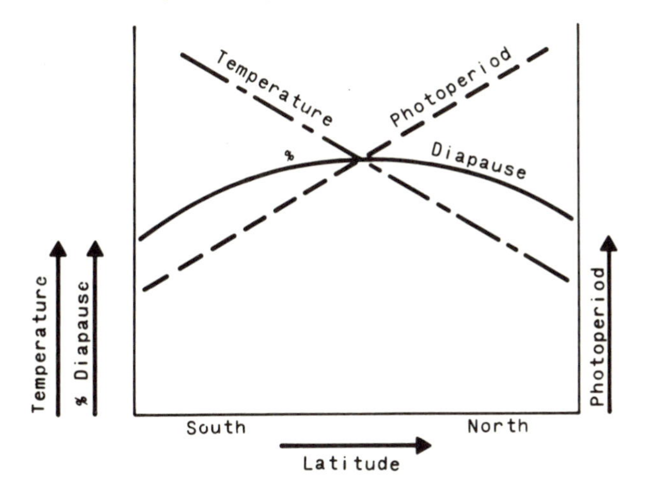

Figure 6-5. Illustration of the photoperiod-temperature interaction and its effect on the diapause response of *Ostrinia nubilalis* (Hbn.) (Showers et al. 1975).

when individual insects in a specific bivoltine or multivoltine population adapt to the relatively long nights and make efficient use of the available warm temperatures, the diapause response during summer would drop again and few would diapause (south portion of Fig. 6-5). Then the southward expansion of ECB is being influenced by summer diapause because of the inability of ECB populations to adjust rapidly to the interaction between photoperiod and temperature (Showers 1979). This interaction forms a barrier at specific latitudes, forces ECB into summer diapause, and impedes further southward movement. The reader should remember, however, that the interaction between photoperiod and temperature is sufficient at any latitude in the United States to place most members of the northern (univoltine) ecotype in summer diapause. Therefore, the interpretation presented in Fig. 6-5 pertains only to bivoltine and multivoltine populations.

3 Manipulation of the Diapause Response

It is conceivable that the diapause response of ECB could be used to the disadvantage of the species. The escape from winter afforded by diapause could in fact prove detrimental to a population of ECB that diapaused during the summer, because the diapausing larvae would be exposed to hazards during the long warm season (Sparks et al. 1966b). Conversely, if a univoltine or bivoltine population of ECB could be kept out of diapause, the resulting adults would be exposed to a different set of hazards during the cold season.

The NC-105 Committee therefore suggested that, if enough of the southern ecotype of ECB could be produced in the laboratory, modified recurrent selection of this ecotype, such as is used by plant breeders to develop and improve plant hybrids, should produce a low-diapausing population. This population, when introduced into populations of northern or central ecotype ECB, would radically reduce the incidence of diapause in these naturally occurring populations. Consequently, many of the individuals would be forced to pupate and emerge as adults during the late autumn and would die with the onset of winter.

During the late summer of 1974, several hundred ECB larvae were collected near Tifton, Georgia, and shipped to the Corn Insects Research Laboratory, Ankeny, Iowa. These larvae were placed on a meridic diet developed and improved by Lewis and Lynch (1969, 1970) and allowed to mate and reproduce. The progeny were reared according to a modification of Reed et al. (1972).

The modified recurrent selection technique used to develop a low-diapausing population of ECB, designated North Central Selection, is presented in Table 6-3. The diapausing photoperiod, 12L:12D, was used with generation I and maintained throughout the selections, but a decreasing thermoperiod was used with each succeeding generation.

The results of the recurrent selections are reported in Table 6-4. Eighty percent of the survivors of generation I diapaused and only 1,720 eggs were available from the 15 females selected to produce generation II (Table 6-3). Again the percentage diapause was high; however, 4,000 eggs were available from the 20 females selected to produce generation III. There was a moderate decrease in percentage diapause for generation III and the number of available females produced about 7,000 eggs. Approximately 6,000 individuals of generation IV pupated and became adults. Only 7% remained larvae and entered diapause. The recurrent selection technique used in this study allowed a rapid selection of nondiapausing traits under a high diapause-inducing regimen.

Another experiment was conducted in the laboratory to determine whether the diapause response of the North Central Selection (NCS) was significantly different from

Table 6-3. Development of North Central Selection (NCS) low-diapausing population of *Ostrinia nubilalis* (Hbn.)

Regimen[a]	
24L/26.7 °C	Population from southern ecotype 300 pairs bulk-mated
	Generation I
12L:12D/17.2-28.3 °C	15 pair selected for fecundity
	Generation II
12L:12D/15.6-26.7 °C	Sibling mate 20 lines with largest moth emergence; 10 lines with largest fecundity saved
	Generation III
12L:12D/14.4-24.4 °C	All possible crosses, except sibling; 100 crosses with largest fecundity saved
	Generation IV
12L:12D/11.1-24.4 °C	Sibling mate 20 lines with largest moth emergence; 10 lines with largest fecundity saved
	Generation V
Same as generation IV	Maintain by bulk matings: Cross to native *O. nubilalis* (NCS female X native male; native female X NCS male)

[a]Photoperiod/thermoperiod.

Table 6-4. Reduction in incidence of diapause for the southern ecotype of *Ostrinia nubilalis* (Hbn.) subjected to modified recurrent selection

Generation	Eggs	Pupae	Diapausing larvae	% Diapause	% Survivorship
I	16,000	2,383	9,617	80	75
II	1,720	315	1,147	78.4	85
III	4,000	1,111	2,240	66.9	84
IV	6,680	5,921	425	7	95
V	5,940	5,435	298	5	95

the diapause response of the parent type from southern Georgia. Neonates from 1,200 eggs of each population were reared singly on a meridic diet in a diapause-inducing regimen of 12L:12D photoperiod and 11.1-24.4 °C thermoperiod (Table 6-5). The survivorship of the two populations was similar, but 51% fewer insects of the selected population diapaused.

A field experiment was conducted in central Iowa to test the hypothesis that an introduction of low-diapausing genes into a central ecotype population of ECB would radically reduce the incidence of diapause, thus seriously affecting the overwintering capabilities of the hybrids. Reciprocal matings were made between fifth-generation NCS and a native (Iowa) population of ECB (Table 6-2). Resulting egg masses from the reciprocals and the parental matings were isolated on 12 silking and tasselling hybrid corn plants (60 eggs per plant) under a 2-m^3 screen field cage. Each treatment was replicated four times (16 field cages). Beginning August 31st, the corn plants and the interior of each cage were observed daily for adult ECB emergence. Each adult was counted and sexed. To determine the percentage diapause, on October 20-21, each plant in each cage was split from tassel to roots and all larvae were collected, isolated and held for three months in a nondiapausing regimen (24-hr light and 28 °C) in the laboratory.

The summer survivorship of the NCS and the native (Iowa) populations and of progeny resulting from reciprocal matings of the two populations of ECB in the field cages is summarized in Table 6-6. Although survivorship of the selected population (NCS) was high in the laboratory (Table 6-4), survivorship of this population was significantly lower than the survivorship of the progeny of the native and one reciprocal mating. Thus, the NCS population is not particularly suited to the summer environment of central Iowa. The data also indicate that the ability to survive may be related to the parental male (NCS × IA vs. IA × NCS, Table 6-6).

Table 6-5. Survivorship and diapause incidence[a] of two populations of *Ostrinia nubilalis* (Hbn.) reared in the laboratory in a diapause-inducing regimen

Population	Numbers of:			Percentage	
	Eggs	Larvae in diapause	Pupae	Survivorship	Diapause
Georgia	1200a	959a	179a	95a	80a
North Central Selection	1200a	349b	820b	97a	29b

[a]Means followed by the same letter, within columns, are not significantly different at the 1% level.

Table 6-6. The survivorship per hybrid corn plant[a] or progeny from parental and reciprocal crosses of *Ostrinia nubilalis* (Hbn.), Ankeny, Iowa, 1975

Mating (female X male)	Replicates				
	I	II	III	IV	Average
NCS X NCS	3.4	4.0	2.2	3.3	3.2a
NCS X IA	12.0	7.3	7.7	12.7	9.9c
IA X NCS	7.5	3.5	3.4	5.9	5.1ab
IA X IA	6.2	7.1	6.2	10.9	7.6bc

[a]Means followed by the same letter are not significantly different at the 5% level, Duncan's Multiple Range Test.

Percentage diapause of four populations of ECB resulting from NCS, a natural population, and their reciprocals is presented in Table 6-7. As expected, the incidence of diapause at latitude 41°43'N (Showers et al. 1975) was nearly complete for the Iowa population. However, when the low-diapausing genes of NCS were introduced into the Iowa population, the incidence of diapause was dramatically affected. Again, as previously reported (Showers et al. 1972, Reed et al. 1981), the diapause responses seems to be sex-linked to the parental male (Table 6-7). However, the fact that the diapause incidence of the Iowa population was high does not mean that this population is homozygous for diapause. If it were, it would diapause at all latitudes as does the northern ecotype (Fig. 6-4). Indeed, the heterozygosity for diapause means that the Iowa male contributes to the relatively large adult emergence, which is expressed as relatively low-percentage diapause for the progeny of the NCS female X IA male mating (Table 6-7).

The recurrent selection process of producing the NCS population of ECB and the resulting field study of this population suggest that the diapause response of populations of ECB that represent the central and northern ecotypes can be manipulated. After the introduction of NCS genes, large numbers of ECB larvae that would normally diapause will presumably pupate, emerge as adults during late September and October, die, and thus reduce the number of available adults for mating and egg-laying the following spring.

Table 6-7. Percentage diapause[a] of an *Ostrinia nubilalis* (Hbn.) population selected for nondiapause, a natural population, and their reciprocals, Ankeny, Iowa, 1975

Mating (female X male)	Replicates				
	I	II	III	IV	Average
NCS X NCS	5.4	2.1	3.8	0	2.9a
NCS X IA	14.4	23.5	17.4	45.7	25.3b
IA X NCS	15.2	7.7	8.1	2.8	8.5a
IA X IA	98.6	93	100	98	97.3c

[a]Means followed by the same letter are not significantly different at the 5% level, Duncan's Multiple Range Test.

4 Discussion

Important to understanding evolution is the study of variation among species and populations within species. The distribution, behavior and population growth of different species are all consequences of genetic and environmental influences on phenology (Tauber and Tauber 1973), fecundity and mortality (Dingle 1972). Insects are excellent animals for the study of these selective factors. Their small size, short generation times, ease of rearing, and particularly their abundance, makes it easy to address evolutionary questions (Dingle 1979). Many insect ecologists are concerned with how natural variation in diapause and migration influences the evolution of life histories (Dingle 1979). Of particular significance is how either of these two escape mechanisms can be manipulated to control populations of pest insects.

Many of the most serious agricultural pest insects have very broad geographical distributions. Therefore, genetic differences must exist within these species as a result of adaptation to local conditions within broader geographic areas. These insects sense environmental cues (e.g., photoperiod) that portend the change in seasons and thus escape adverse periods through diapause, migration, or a combination of the two. The idea of suppressing insect populations adapted to particular climates by genetic manipulation has been suggested by Hogan (1966), Masaki (1968) and Klassen et al. (1970b).

Klassen et al. (1970a) suggested that conditional lethal traits such as inability to diapause, inappropriate critical thresholds of diapause-inducing cues (photoperiod, temperature and diet), and inability to develop cold-hardiness, as well as other adaptations, could be selected for and used to suppress a pest insect. However, in all of their calculations a dominant sex-linked gene, dominant autosomal gene, or three or four autosomal genes with additive effects were necessary to promote the conditional lethal traits.

Studies conducted by Sparks et al. (1966a), Chiang et al. (1968) and Showers et al. (1972) suggested that the diapause response of *O. nubilalis* seemed dominant under certain environmental conditions and recessive under other environmental conditions. Showers et al. (1975) addressed this confusing issue, and found that certain interactions between photoperiod and temperature allowed the diapause condition to be enhanced and seemingly dominant. When daylength was long, the effects of cool temperatures (in the north) were suppressed and populations failed to diapause. Conversely, if temperature was warm, the effects of short daylength (in the south) were overridden, and the diapause response was suppressed and seemingly recessive (Fig. 6-5).

Showers et al. (1972) suggested that the diapause response of ECB is sex-linked to the parental male. This possibility is again evident when the results in Table 6-7 are examined. Reed et al. (1981) addressed this problem and concluded that diapause in ECB is controlled genetically by a system that has a pattern substantially consistent with sex-linked inheritance. However, they also concluded that the sex-linked gene was nondominant and not automsomal.

The conclusions expressed by Reed et al. (1981) conflict with the requirements necessary to promote conditional lethal traits calculated by Klassen et al. (1970a). Nevertheless, the results presented here indicate that the diapause response can indeed be selected against. This enhances the possibility that populations inhabiting areas where diapause is essential for survival can be suppressed by replacing them with a strain carrying a genetic factor which prevents diapause.

I must caution, however, that inundating an area once with a strain of ECB selected for low diapause will do little to suppress the native population. These low-diapausing genes will become randomly distributed throughout the population by the F_3 generation. Inundation each spring (when the native population is small) over a 3-4-year period might result in a shift of high-diapausing northern or central ecotypes to low-diapausing populations that could not survive the cold climate.

Since 1975, the NCS (low-diapausing) population of ECB has been used in reciprocal crosses and backcrosses at field locations throughout the distribution of ECB in the United States. These studies are being conducted by the NC-105, North Central Regional Technical Committee, and the results will be reported elsewhere.

Acknowledgments. Journal Paper J-9861 of the Iowa Agricultural and Home Economics Experiment Station, Ames, Project 2419. Contribution from North Central Project NC-105.

References

Apple, J. W.: Corn borer development and control on canning corn in relation to temperature accumulation. J. Econ. Entomol. 45, 877-879 (1952).

Arbuthnot, K. D.: Temperature and precipitation in relation to the number of generations of European corn borer in the United States. USDA Tech. Bull. 987 (1949).

Beck, S. D.: Photoperiodic induction of diapause in an insect. Biol. Bull. 122, 1-12 (1962).

Beck, S. D.: Insect Photoperiodism. New York: Academic Press, 1968.

Beck, S. D., Apple, J. W.: Effects of temperature and photoperiod in voltinism of geographical populations of the European corn borer, *Pyrausta nubilalis.* J. Econ. Entomol. 54, 550-558 (1961).

Caffrey, D. J., Worthley, L. H.: A progress report on the investigations of the European corn borer. USDA Bull. 1476 (1927).

Chiang, H. C., Keaster, A. J., Reed, G. L.: Differences in ecological responses of three biotypes of *Ostrinia nubilalis* from the north central United States. Ann. Entomol. Soc. Am. 61, 140-146 (1968).

Chippendale, G. M.: Hormonal regulation of larval diapause. Ann. Rev. Entomol. 22, 121-138 (1977).

Dingle, H.: Migration strategies of insects. Science 175, 1327-1335 (1972).

Dingle, H.: Adaptive variation in the evolution of insect migration. In: Movement of Highly Mobile Insects: Concepts and Methodology in Research. Rabb, R. L., Kennedy, G. G. (eds.). Raleigh, N.C.: North Carolina State Univ. Press, 1979.

Felt, E. P.: The European corn borer. Cornell Ext. Bull. 31, 35-48 (1919).

Harris, H. M., Brindley, J. M.: The European corn borer in Iowa, J. Econ. Entomol. 35, 940-941 (1942).

Hogan, T. W.: Physiological differences between races of *Teleogryllus commodus* (Walker) (Orthoptera: Gryllidae) related to a proposed genetic approach to control. Aust. J. Zool. 14, 245-251 (1966).

Klassen, W., Creech, J. F., Bell, R. A.: The potential for genetic suppression of insect populations by their adaptations to climate. ARS, USDA Misc. Publ. 1178 (1970a).

Klassen, W., Knipling, E. F., McGuire, J. U.: The potential for insect population suppression by dominant conditional lethal traits. Ann. Entomol. Soc. Am. 63, 238-255 (1970b).

Lewis, L. C., Lynch, R. E.: Rearing the European corn borer on corn leaf and wheat germ diets. Iowa State J. Sci. 44, 9-14 (1969).

Lewis, L. C., Lynch, R. E.: Treatment of *Ostrinia nubilalis* larvae with Fumidil B. to control infections caused by *Perezia pyraustae*. J. Invertebr. Pathol. 15, 43-48 (1970).

Masaki, S.: Geographic adaptation in the seasonal life cycle of *Mamestra brassicae* (L.) (Lepidoptera: Noctuidae). Hirosaki Univ. Facul. Agric. Bull. 14, 16-26 (1968).

Mutchmor, J. A., Beckel, W. E.: Some factors affecting diapause in the European corn borer, *Ostrinia nubilalis* (Hbn.) (Lepidoptera: Pyralidae). Can. J. Zool. 37, 161-168 (1959).

North Central Regional Committee, NC-105: The European corn borer and its control in the North Central States. No. Cent. Reg. Publ. 22. IA. Agric. and Home Econ. Exp. Stat. Pamph. 176 Rev. (1972).

Reed, G. L., Showers, W. B., Huggans, J. L., Carter, S. W.: Improved procedures for mass rearing European corn borer. J. Econ. Entomol. 65, 1472-1476 (1972).

Reed, G. L., Guthrie, W. D., Showers, W. B., Barry, B. D., Cox, D. F.: Sex-linked inheritance in diapause of the European corn borer and its significance to diapause physiology. Ann. Entomol. Soc. Am. 74, 1-8 (1981).

Showers, W. B.: Effect of diapause on the migration of the European corn borer into the southeastern United States. In: Movement of Highly Mobile Insects: Concepts and Methodology in Research. Rabb, R. L., Kennedy, G. G. (eds.). Raleigh, N.C.: North Carolina State Univ. Press, 1979.

Showers, W. B., Reed, G. L.: Three generations of European corn borer in central Iowa. Proc. N. Cent. Br. Entomol. Soc. Am. 26, 53-56 (1971).

Showers, W. B., Brindley, T. A., Reed, G. L.: Survival and diapause characteristics of hybrids of three geographical races of the European corn borer. Ann. Entomol. Soc. Am. 65, 450-457 (1972).

Showers, W. B., Chiang, H. C., Keaster, A. J., Hill, R. E., Reed, G. L., Sparks, A. N., Musick, G. J.: Ecotypes of the European corn borer in North America. Environ. Entomol. 4, 753-760 (1975).

Skopic, S. D., Bowen, M. F.: Insect photoperiodism: An hourglass measures photoperiod time in *Ostrinia nubilalis*. J. Comp. Physiol. 111, 249-259 (1976).

Sparks, A. N., Brindley, T. A., Penny, N. D.: Laboratory and field studies of F_1 progenies from reciprocal matings of biotypes of the European corn borer. J. Econ. Entomol. 59, 915-921 (1966a).

Sparks, A. N., Chiang, H. C., Keaster, A. J., Fairchild, M. L., Brindley, T. A.: Field studies of European corn borer biotypes in the midwest. J. Econ. Entomol. 59, 922-928 (1966b).

Srb, A. M., Owen, R. D., Edgar, R. S.: General Genetics, 2nd edn. San Francisco: W. H. Freeman, 1965.

Tauber, M. J., Tauber, C. A.: Insect phenology: Criteria for analyzing dormancy and for forecasting postdiapause development and reproduction in the field. Search (Agric.), Cornell Univ. Agric. Exp. Stn. 3, 1-16 (1973).

Vinal, S. C.: The European corn borer, a recently established pest in Massachusetts. Mass. Agric. Exp. Stat. Bull. 178, 147-152 (1917).

Chapter 7

Natural Selection and Life History Variation: Theory plus Lessons from a Mosquito

CONRAD A. ISTOCK

1 Introduction

In theory, life history patterns and polygenic variation in fitness characters provide the selection regimes and essential raw material for microevolution. Over the geographical range of a species, fixed, genetic differences in life history patterns are the result of past microevolution. In any local population of a species, the dynamic interrelations of genetic variation, life history phenomena and natural selection may be observable. For longer-term geographical patterns to continuously or recurrently change, there must persist local genetic variability for at least some of the fitness characters, and there is now considerable evidence that this type of genetic variation, either expressed or potential, is common in natural populations (Istock 1981). Geographical differentiation within a species implies past directional selection, while the conservation of genetic variability in local populations implies the predominance of some form of stabilizing selection. In this paper I will present theoretical and empirical results from studies with the pitcher-plant mosquito, *Wyeomyia smithii,* which expose both geographical and local genetic variation and something of the dynamic processes of stabilizing selection occurring at the local level.

At the fundamental level of definition captured in the Lotka-Euler equation, the primary fitness and life history characters are synonomous (Hairston et al. 1970). These primary characters are: survival probability, fertility and development time. Further, the genetic variation in these characters defines limits to the future patterns of life history achievable by natural selection (Stearns 1976). A complete theory of life history evolution would be able to predict such future transformations. We are probably still a long way from that goal, but general aspects of such a theory emerge from the results of studies with *W. smithii* included in this paper: (1) we can detect local genetic variation for life history characters and see the kind of stabilizing selection taking place, and (2) we can reconstruct in part how this detectable variation in local populations has been converted to geographical differentiation.

The Lotka-Euler representation of fitness and life history characters will, however, not suffice to capture the dynamics of the joint processes of population replacement and natural selection as they appear in a natural population of *W. smithii* at Kennedy Bog in western New York. Here seasonal uncertainty enters in an important way, and genetic variation and selection upon two correlated development characters, larval development time and tendency to diapause during the warm season, assume central significance. The pattern of natural selection detected is one of fluctuating-stabilizing selection arising from environmental uncertainty. An appropriate measure of fitness in this setting is not the intrinsic rate of increase of the Lotka-Euler expression, but the cumulative product of sequential net reproductive rates, R, for a matrilineally defined lineage ("the unit of selection" in this instance) passing through indefinite time. This existential definition of fitness was original with Thoday (1953, 1958), who expressed it as the probability of long-term persistence of a population or population subunit. Thoday's idea is not only conceptually interesting, but contrary to criticism (Ayala 1965) has actually been applied directly in laboratory population experiments (Saul 1970). At this stage in the development of microevolutionary theory it is safe to say that no single "unit of selection" or definition of "fitness" suffices, but the reasons for choosing the matrilineage and cumulative R definition in this instance should become clear below.

2 Background from Previous Studies with *Wyeomyia smithii* Populations

(1) Geographical variation in photoperiodically controlled diapause in *W. smithii* shows a smooth cline from Alabama, near the southern limit of species range, to Massachusetts, and presumably all the way on up to the northern limit of the species range. This cline is beautifully demonstrated in the work of Bradshaw (1971, 1976) and Bradshaw and Lounibos (1972, 1977). These papers not only reveal the phenotypic cline, but through a series of laboratory crosses show a genetic basis for the cline. Figure 7-1 provides new data from my own studies demonstrating that the cline is phenotypically continuous over a smaller, north-south distance from southern Michigan to the south shore of Lake Superior.

(2) The fraction of a cohort or population of *W. smithii* entering diapause in response to a particular photoperiod is strongly altered by food concentration. Near the critical photoperiod (50% diapause expected), high food concentration reduced the fraction diapausing to as low as 0.15 and low food concentration brought on 100% diapause (Istock et al. 1975). This indicates that *W. smithii* larvae will respond by diapausing or not diapausing depending on the degree to which the population is under density-dependent or density-independent control with respect to food. We have good evidence from field experiments that periods of both kinds of population control typically occur in the same season (Istock et al. 1976b). Hence, information about environmental quality may be received by larval monitoring of the food supply and environmental uncertainty for the larvae thereby reduced. However, the diapausing instar, in summer or winter, in this species is the penultimate, or third, larval stage. Upon molting to the fourth instar, individuals of northern *W. smithii* make an essentially irrevocable and blind commitment to adulthood, a stage in which they cannot survive the winter. Hence, the effect of environmental variation and uncertainty falls most heavily on the fourth instar, pupal and adult stages. The effect of food concentration

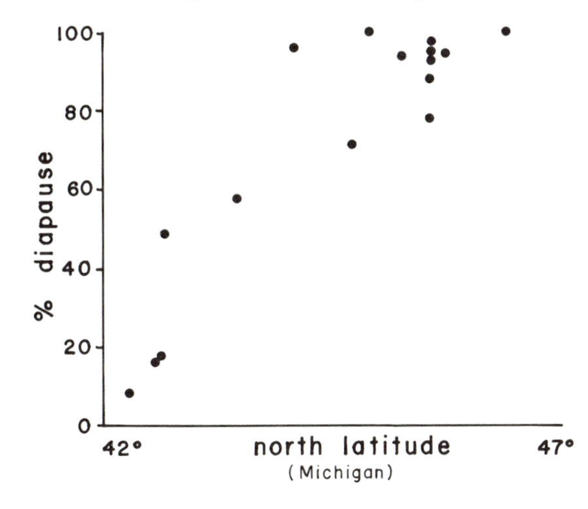

Figure 7-1. Percentage of third-instar diapause expressed by stocks of *W. smithii* obtained from natural populations along a transect from southern Michigan (Ann Arbor) to the south shore of Lake Superior (Grand Marais). Each stock was tested in the same laboratory environment with a photoperiod of 15 hr light and a daily temperature cycle of 15-30 °C. The values for % diapause become crowded at the top of the graph because the experimental photoperiod is substantially below the critical daylength (50% diapause expected) for the more northern populations in the series. Sample sizes of approximately 100 larvae for each locality were used.

on the diapause vs. nondiapause decision has assumed new meaning with the discovery that the fourth instar of *W. smithii* is the period when all storage for egg production (fertility) takes place (there is no bloodfeeding in this mosquito). There is also a flexibility to the duration of the fourth instar such that a longer duration allows partial compensation in egg production when food is at a lower level. Adults take nectar for physiological maintenance, but nectar makes no contribution to egg production. Furthermore, the materials stored for egg production in the fourth instar cannot be drawn upon for adult maintenance (Lang 1978, Moeur and Istock 1980).

The next series of conclusions about local populations of northern *W. smithii* are summarized to a large extent in Figure 7-2.

(3) Laboratory studies employing offspring-parent regression and artificial selection have revealed polygenic variation for development time and diapause in a *W. smithii* population. That the two characters are genetically correlated was strikingly shown by mass reciprocal crossing of the selected populations. This genetic substructure for two aspects of development time codes for a range of phenotypes from extreme fast-developing and diapause-resisting to extreme diapause-prone and slow-developing. When mass selection for either the "fast" or "diapause" extreme is performed, the additive variance for development time and its heritability decline to zero. When the fast and diapause stocks obtained by artificial selection are crossed, intermediates are obtained (Istock et al. 1976a, Istock 1978). We have repeated these results often enough to be sure that the polygenic variation underlying these two characters is a persistent property of the natural population. In an unselected population the phenotypic variation is completely continuous. Hence the life history variation which emerges must produce a

range of patterns extending from near univoltinism to multivoltinism of three generations per season at Kennedy Bog (Istock 1978).

The consequent range of variation in potential, cumulative net reproductive rates for a season, R(s), will be approximately 10-1000 between the two extreme patterns. With mixing of diapausers and nondiapausers during a season an effective number of generations per season, x, is created, and x will be a continuous variable controlling the expressed distribution of R(s) (Fig. 7-2). In the jargon of evolutionary ecology, this looks like a bet-hedging or mixed strategy phenomenon (Stearns 1976), but one for which the genetic basis has been demonstrated. Thus, developmental variation controls the timing and number of generations per year, stabilizing selection is required to preserve the underlying genetic variation, and sexuality functions to deploy the bet-hedging tactic (Istock 1978).

(4) One question which arises immediately is whether or not individual females really deploy the bet-hedging tactic. Is the developmental variation, particularly that

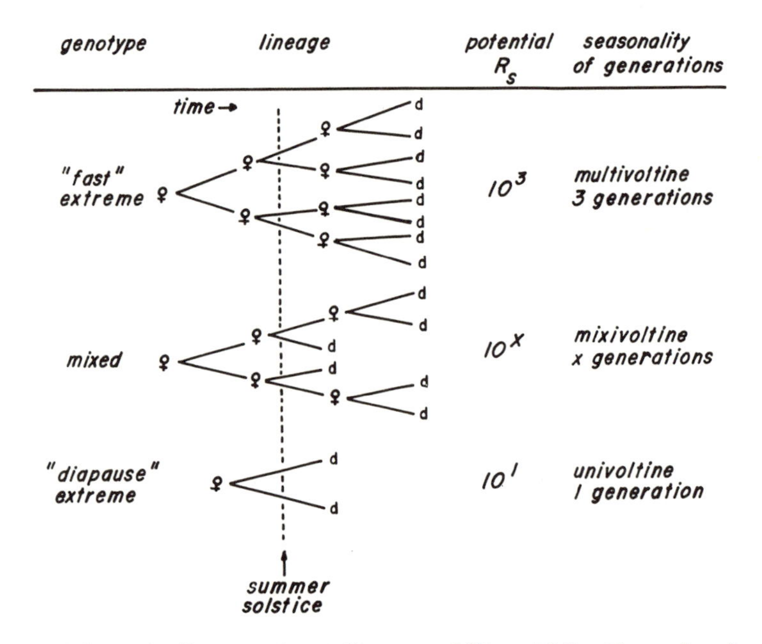

Figure 7-2. Schematic diagram of matrilineages of *W. smithii* with varying degrees of diapause during the warm season, as they would occur at Kennedy Bog near Rochester, New York. Fast, mixed and diapause refer to genotypic states isolated in laboratory studies of the quantitative inheritance of diapause using populations of *W. smithii* colonized from Kennedy Bog. A net reproductive rate per generation of 10, based on laboratory estimates, is used for illustrative calculations of whole-season, cumulative net reproductive rates, R(s), for each lineage. For simplicity, dichotomous branching at each reproductive node (♀) of a lineage is shown, rather than the polychotomous branching which actually occurs in *W. smithii* and most higher organisms. The d refers to diapause females within a lineage. Between the diapause and fast extremes any effective generation number $1 < X < 3$, can occur due to the mixing of diapause and reproductive females. This variable x is the "mixivoltine" condition. Further discussion in text.

for diapause, subject to individual selection, or do we need to invoke group selection at the outset to explain the persistence of variation for lower potential R and R(s) for some individuals? In the laboratory we measured the diapause fraction for progeny larvae from single wild-caught females. The larvae were grown under standard conditions (Istock et al. 1975, 1976a) with the photoperiod near the critical value for 50% diapause, mimicking conditions of early or late summer. The tests were run with progeny from females taken throughout the 1977 season at Kennedy Bog, and the results appear in Fig. 7-3. As larvae, the wild-caught females had experienced the prevailing, natural conditions of larval growth preceding their capture as pupae or adults. Almost the entire range from all developers to all diapausers was observed among these groups of sibs early in the summer. The fraction of diapause within families declines through the season, because individuals with high gene dosage for diapause become diapausers themselves and remain in the pitchers as third-instar larvae prepared for winter. Even so, there is still a tendency for the mixed strategy to predominate in late summer. The data strongly suggest that the bet-hedging tactic occurs in nature, and that it is expressed among the progeny of single females (and possibly single males as well, because laboratory tests so far indicate single mating in *W. smithii*). The varying dosages of diapause genes, carried by these females and their unknown mates, almost certainly come about through genetic segregation and recombination mediated by sexuality. Earlier results showed an aggregate, seasonal pattern to the expression of diapause in nature which is in accordance with these results for single females (Istock 1978).

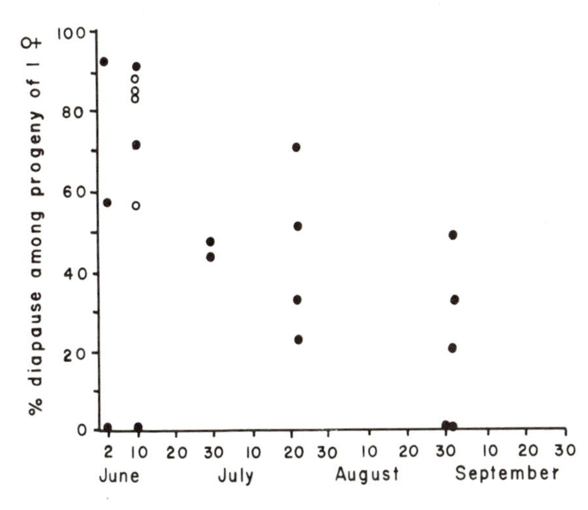

Figure 7-3. Percent diapause expressed by progeny of single wild-caught females of *W. smithii* from the Kennedy Bog population (closed circles) and the Zurich Bog population (near Rochester, New York, open circles) during the summer of 1977. Further explanation in text.

3 Geographical Patterns in the Genetics of Life History

Genetically based geographical diversification may be the prelude to speciation. With reproductive, fitness and life history characters so much interrelated, it would seem that genetic divergence in these characters would be the *sine qua non* of incipient speciation, particularly if diverging populations are geographically isolated. Conversely, divergence in superficial characters may have nothing to do with speciation if there is no developing reproductive isolation at the more fundamental level of the fitness characters. In fact, southern and northern populations of *W. smithii* were once named as distinct species on morphological grounds (Bradshaw and Lounibos 1977). Northern and southern populations have probably not exchanged genes for about 20,000 generations. It thus becomes interesting to ask if the southern and northern populations carry different genetic systems for life history pattern, and whether these genetic systems show deleterious effects on the fitness characters once they are recombined experimentally in the laboratory. Thus far, the answer to the first question is yes and to the second question no, as shown in Tables 7-1 and 7-2.

Tables 7-1 and 7-2 provide data for some fitness characters for the f1 progeny from reciprocal crosses of stocks taken from northern Ontario, Canada (65 km due north of Sault Saint Marie) and northwest Florida (Florida panhandle). The progeny from these crosses were grown in conditions roughly simulating summer photoperiod and temperature conditions of the respective localities from which they originated (descriptions of the simulated environments appear in Table 7-1). First of all there is no sign of incompatibility for the systems controlling development or survival. The same was true

Table 7-1. F1 progeny from crosses between geographically divergent populations of *Wyeomyia smithii*: development time and diapause

Cross	Dev. time (± SE)	% Diapause	Env.	N
Fla X Fla	27.46 (0.445)	12.26	Fla	136
Fla X Ont	26.52 (0.667)	29.71	Fla	123
Fla X Ont	19.00 (0.232)	1.09	Fla	181
Ont X Fla	20.13 (0.789)	32.86	Fla	47
Ont X Fla	18.46 (0.597)	27.27	Fla	32
Ont X Ont	28.00 (1.983)	96.27	Fla	6
Ont X Ont	31.22 (0.827)	53.93	Ont	41
Ont X Fla	21.18 (0.339)	0.00	Ont	40
Fla X Ont	20.81 (0.185)	1.11	Ont	178
Fla X Fla	27.56 (0.419)	6.78	Ont	110

Each cross is arranged as female parent source x male parent source. All were mass crosses using 10 females and 10 males.
Dev. time = the number of days elapsed from egg to pupa on the average.
Fla. env. = diurnal temp. cycle of 23-32 °C, 14 hr light per day.
Ont. env. = 11-24 °C cycle, 16 hr light per day.
N = the number of nondiapausing larvae, i.e., those used to calculate mean development time.

for egg production. Hence no deleterious effects on the primary fitness characters appeared in the f1 generation. The f2 and subsequent generations of hybrids and controls were equally vigorous.

Turning to genetic effects on life history patterns, there are some clear indications of geographical divergence and specialization. When Ont X Ont (Ontario) progeny grew in the mock Ontario environment, they revealed that the photoperiod used (16 hr light) is close to their critical photoperiod because 53.93% of the larvae entered diapause. Fla X Fla progeny show low diapause under the long daylength of the Ontario environment; the fact that there is almost 7% diapause is probably due to the low temperature of this environment.

Surprisingly, the progeny of the reciprocal crosses between localities show an effect akin to nearly total dominance in the suppression of diapause by the Florida side of the cross. A similar surprise appears in the effects on development time all through the data of Table 7-1, except that the result looks more like overdominance. From the mortality data of Table 7-2 there is also a suggestion of overdominance. In the mock Florida environment the Fla X Fla progeny show much less than 50% diapause, because the photoperiod used (14 hr light) is well above the critical photoperiod for the Florida latitude (Bradshaw and Lounibos 1977, their Fig. 8). A longer than natural photoperiod was used for the Florida environment in an attempt to obtain some nondiapause from the Ontario stock. Nevertheless, the short photoperiod of the Florida environment suppressed nearly all completion of the life cycle for the Ont X Ont progeny. A dominance-like effect upon diapause does not occur for the progeny of reciprocal crosses except in one, perhaps anomalous, case. In the other three cases the diapause fraction is intermediate.

Taken together, the data of Table 7-1 suggest that there has been considerable evolutionary specialization and divergence in the life history patterns underlying genetic

Table 7-2. F1 progeny from crosses between geographically divergent populations of *Wyeomyia smithii:* mortality rates

Cross	% Larval mortality	Env.
Fla X Fla	21.32	Fla
Fla X Ont	5.91	Fla
Fla X Ont	3.68	Fla
Ont X Fla	5.41	Fla
Ont X Fla	4.35	Fla
Ont X Ont	7.47	Fla
Ont X Ont	11.00	Ont
Ont X Fla	0.00	Ont
Fla X Ont	5.76	Ont
Fla X Fla	15.11	Ont

The sample sizes for each cross are the same as the respective total N (diapause and nondiapause larvae) for Table 7-1.

systems of latitudinally separate populations of *W. smithii.* Furthermore, the interactions of these divergent systems lead to complex patterns of life history expression in hybrid populations, patterns which are sometimes strongly responsive to environmental conditions. It seems likely that synthetic populations once formed in the laboratory might quickly readopt the observed geographical specializations of their source localities if left in the appropriate environment. No experiment of this type has been tried yet.

Another important observation which can be made from Table 7-1 is that local variation for development time and diapause exists in both northern and southern populations. Presumably, this local variation is partly genetic over much of the species range, just as it is at Kennedy Bog. It will be interesting to find out whether *W. smithii* remains genetically variable for these traits at the northern limit of its range.

4 A Diapause Model for *Wyeomyia smithii*

Now I want to focus on the adaptive significance of local, genetic variation in the tendency to diapause. It is obvious that a lineage will vanish if all of its constituent individuals are reproductives in a generation when the net reproductive rate, R, is zero. Assume that we are considering the same parts of the season when successful reproduction will occur in better years. Conversely, in such a generation a lineage with some or all diapause individuals will survive to the next generation, with an effective net replacement rate dependent on its diapause fraction and the probability, s, of one generation survival in diapause. We have observed an instance when the net reproductive rate of the third generation at Kennedy Bog was effectively zero (Istock 1978). Even when zeros do not occur, lineages with some diapausers will have higher realized fitness in any generation in which realized R falls below the probability of survival in diapause. Thus the repeated production of diapausers becomes a hedge against one or more low values of R and above all against R=0. Fluctuation of R to values lower than s represents major environmental uncertainty in a species, like *W. smithii,* in which individuals must commit themselves to reproduction followed unalterably by death, and must make this commitment in the absence of reliable information about the future quality of the environment in which they must reproduce.

The photoperiod can tell larvae where they are in the solar cycle of the seasons, but it carries no information about transient year-to-year fluctuation of important proximate environmental variables. In the north, at least, a bog may nearly dry up in midsummer, and it may be highly unpredictable in its weather at the beginning and the end of the season. Selection will favor an intermediate or mixed strategy if the fast extreme is periodically selected against by poor conditions in the first or third generation, to use the Kennedy example, and if at least some lineages have the effective number of generations per season greater than one, i.e., the pure diapause extreme is usually less fit (Istock 1978).

Imagine a fluctuating, positively autocorrelated environment in which the concentration of genes for diapause waxes and wanes. Sexuality will on average increase the tendency to diapause for all individuals and lineages when high diapause is favored, and lower the frequencies of genes for diapause following periods favorable to immediate reproduction. The individual at each branching of a lineage can either enter diapause or attempt to reproduce, but a lineage can have any mixture of the two options, and a system of lineages with genetic exchange can track the uncertain environment: hence the need to consider the lineage or some level of population organization above the level of an individual. The problem before us is to incorporate the genetic, demographic and environmental parts of the process into a single model.

Consider a population composed of L matrilineages defined by a set of female founders which were contemporaneous at some arbitrary time in the past. By definition each individual in the population in any subsequent generation t can be a member of only one lineage. Assume that for each female there is one male of identical autosomal genotype and identical phenotype with respect to the tendency to diapause itself, and thus the tendency to transmit genes for diapause to progeny. At each generation t for the ith lineage there are n_t nondiapause females coming from the just previous generation, sd_{t-1} females coming from births in generation t-1 and having been in diapause during the just previous generation, and d_t females entering diapause in the current generation (Fig. 7-4). Thus the total number of females in the ith lineage is:

$$N_t = n_t + sd_{t-1} + d_t,$$

where $2N_t$ is the total size of the lineage, s the constant probablity of survival in diapause for one generation, and $m_t = n_t + sd_{t-1}$ the total number of currently reproductive females within the lineage. As defined here, the model only allows a one-generation diapause, but it is not difficult to modify it so that fractions of each d_t diapause for 1, 2 or more generations.

At each generation the subdivision of progeny into n_t and d_t is determined by the expressed tendency to diapause for the whole lineage in the previous generation, name-

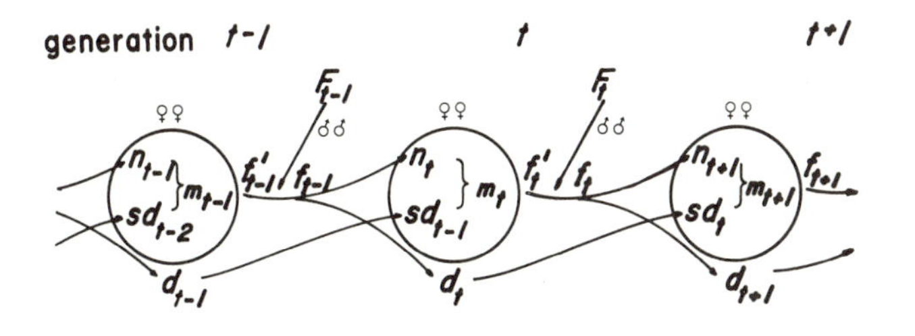

$$N_{t+1} = N_t\left[R_t - \hat{f}_t(R_t - s)\right]$$

Figure 7-4. Diagram of the quantities defined and their temporal relationships in the diapause model for a population of *W. smithii* lineages. One lineage is depicted. Each circle encloses the breeding females for one generation. Diapausing females are shown outside the circle as d. Males enter via panmictic mating, indicated by the downward slanting arrow. The expression at the bottom is the discrete generation recursion equation for one lineage. The mathematical notation is explained in the text.

ly f_{t-1}. Prior to reproduction in the current generation f'_t is the true genetic tendency to diapause carried by the m_t individuals about to produce (Fig. 7-4). Two processes ensue between f'_t and f_t. The first process is panmictic mating, in which males carry in the whole population (all L-lineages together) tendency to diapause F_t, and the second process is the expression of the new tendency to diapause after allowance for a heritability of diapause between 0 and 1, i.e., after the environmental variance and other disturbances to perfect inheritance are allowed to enter. The expression for f'_t is:

$$f'_t = \frac{n_t f'_{t-1} + sd_{t-1} f'_{t-2}}{m_t}$$

and the expression for F_t is:

$$F_t = \frac{1}{\sum_{i=1}^{L} m_{ti}} \sum_{i=1}^{L} m_{ti} f'_{ti}$$

Both are just weighted averages.

We note that both the frequency with which diapause and nondiapause genes are involved in matings, and the average phenotypic values of these genes, are specified by f'_t and the F_t. To describe the average results of panmictic mating between any lineage and the population consider the frequencies p_i and the phenotypic scores x_i associated with four types of matings. The lineage is thus represented by an average female with its tendency to produce diapause and nondiapause individuals in the proportions f'_t and $(1-f'_t)$ respectively. Similarly, all potential males for the population side of each mating have their average surrogate with proportions F_t and $(1-F'_t)$ respectively. This formulation is analogous to thinking of a life table or fertility table as the vital statistics for an average female (Fisher 1958). Now if we assume that an additive genetic system underlies diapause inheritance and hence f'_t and F_t we can write the following:

Mating type lineage/population	Mating frequency	Phenotype
1. diapause/diapause	$p_1 = f'_t F_t$	$x_1 = (f'_t + F_t)/2$
2. nondiapause/diapause	$p_2 = (1-f'_t)F_t$	$x_2 = ((1-f'_t) + F_t)/2$
3. diapause/nondiapause	$p_3 = f'_t(1-F_t)$	$x_3 = (f'_t + (1-F_t))/2$
4. nondiapause/nondiapause	$p_4 = (1-f'_t)(1-F_t)$	$x_4 = ((1-f'_t) + (1-F_t))/2$

where "phenotype" is the expected value due simply to the additive genetic process specifying the phenotype.

Now f_t after mating and the admittance of environmental and other influences on genetic expression can be written as:

$$f_t = r(1-h) + \sum_{i=1}^{4} p_i x_i$$

with the summation containing the estimate of the expected value of f_t in the progeny, h the heritability of diapause, and r a random variable which disturbs the ultimate phenotypic expression as errors of sampling, assortative mating, environmental change or mutation might.

The population of lineages exists in a continually fluctuating and probably autocorrelated environment such that a sequence of average net reproductive rates $\bar{R}_t$ are imposed on each lineage, one new value of $\bar{R}$ each generation. Winter is ignored by assuming that all surviving larvae at the onset of winter survive in diapause through to the next time for reproduction. Genes other than those for diapause are assumed to be randomized within and between lineages. If $\hat{f}_t = d_t/N_t$ is the actual frequency of diapausers for the ith lineage, and $(1-\hat{f}_t) = m_t/N_t$, the discrete generation growth or decline in numbers for any lineage is:

$$\begin{aligned} N_{t+1} &= \bar{R}_t m_t + s d_t \\ &= N_t [\bar{R}_t \frac{m_t}{N_t} + s \frac{d_t}{N_t}] \\ &= N_t [\bar{R}_t (1-\hat{f}_t) + s\hat{f}_t] \\ &= N_t [\bar{R}_t - \hat{f}_t (\bar{R}_t - s)] \end{aligned}$$

from which we see the interplay of the tendency to diapause, survival in diapause, and the environmental fluctuation brought in by a sequency of $\bar{R}$ values. To see this interplay in a simple way note the consequences of high vs. low f_t when $\bar{R}_t$ is greater than 1 (s must be less than or equal to 1) as opposed to the consequences when $\bar{R}_t = 0$. The cumulative net reproductive rate of any lineage would be the cumulative product of the term in brackets. This expression for growth of a lineage with diapause is identical to the model of Takahashi (1977) for a population with diapause, although the derivation is somewhat different. Takahashi also considered diapause as an adaptation to a fluctuating or "unstable" environment.

The most intriguing feature which *W. smithii* suggests for the model is the persistence of genetic variation for diapause. There is also evidence for peristent heritable variation in diapause from studies with other insects including the webworm, *Hyphantria cunea* (Morris and Fulton 1970, Morris 1971), *Drosophila littoralis* (Oikarinen and Lumme 1979), the pink bollworm, *Heliothis zea* (Herzog and Philips 1976), the milkweed bug, *Oncopeltis fasciatus* (Dingle et al. 1977), and the gypsy moth, *Lymantria dispar* (Lynch and Hoy 1978).

It must be recognized that I have assumed that genetic variation for diapause will be conserved under natural selection because that is what the current data indicates is happening in nature. Thus, the expression for each successive value of f_t forces the genetic variation to be reassorted among lineages. The total amount of diapause over all lineages and the underlying concentration of genes for diapause may rise and fall, but the diminution of genetic variability is never enough to prevent at least some lineages from expressing any value of f_t in the range 0 to 1. Maynard Smith (1979) has argued that simple, constant-environment, stabilizing selection will diminish the genetic variation with time, but Lande (1976) has argued on theoretical grounds that recurrent mutation may be generally sufficient to prevent loss of variability.

Preliminary simulations of the diapause model suggest that the model is reasonable, at least to the extent that it preserves an initial set of lineages in an autocorrelated environmental sequence where $\overline{R}$ fluctuates between 0 and 2 (0 rare and never two in a row), with its arithmetic mean over time slightly greater than one. The numerical behavior is complex, but not at all chaotic, and the effect of diapause in buffering against sequences of low $\overline{R}$ is apparent. Further numerical simulations need to be done using a variety of environmental sequences. With only a one-generation diapause no lineage will survive two consecutive values of $R=0$. This suggests that a version of the model with longer durations of diapause for some individuals may be of more general interest. The model in that form might also describe the adaptive function of prolonged and variable seed-dormancy in plants.

The model presented here has some similarities to Cohen's (1970) model for the optimal timing of diapause, though the two models address quite different questions. Cohen included age structure, focused on adult diapause, and assumed that adults receive information about the quality of the future environment prior to the decision between reproduction or diapause. Cohen did not incorporate any genetic structure in his model, which raises the general question of whether we can ever hope to draw solid conclusions about the evolution of life history patterns from models which contain only ecological considerations.

5 Conclusions

Populations of *Wyeomyia smithii* appear to undergo almost continuous natural selection for a shifting and intermediate amount of larval diapause. The pattern of selection may be described as fluctuating-stabilizing selection. Polygenic variation for development time and diapause seem to provide a genetic basis for this type of continual adaptation to an uncertain environment, without steady loss of the requisite genetic variation. Artificial directional selection for either fast, nondiapause development on the one hand, or for diapause on the other hand, in laboratory populations does erode the genetic variation for development time and diapause.

A complete model incorporating genetic and phenotypic variation for diapause, and natural selection in a fluctuating environment, can be constructed to describe the type of fluctuating-stabilizing selection thought to occur in *W. smithii* populations. This form of microevolution is a special case of natural selection for an intermediate optimum (Wright 1968). Under such a model the voltinism of the population becomes a continuous variable. The mode of adaptation postulated accords with the usual notions about a bet-hedging tactic or mixed strategy postulated in evolutionary ecology, except that a genetic basis for such adaptation has been uncovered in a natural population of *W. smithii* and incorporated into the diapause model.

Most of the insect species examined to date appear to express genetic variation for some aspect of developmental timing, at least in the laboratory. The life stage at which development is genetically variable differs from species to species. Frequently, there is heritable variation for a diapause characteristic. In several cases heritable variation for diapause is known to be expressed in nature (Istock 1981). Heritable variation in development time and diapause could provide a generalized adaptation to environmental uncertainties created by human efforts at insect pest control. Pest species might obtain a form of ecological resistance by evolving life history patterns with greater and greater asynchrony of development up to the limits allowed by underlying genetic systems. Non-pest species could undergo similar protective responses. If this kind of generalized life history evolution does occur, it may make spraying or other kinds of control methods increasingly frequent and expensive, all the while buying time for the target species to achieve a physiological or morphological (barrier-to-penetration) form of resistance. This notion is purely speculative, but worth looking for.

Latitudinally arrayed populations of *W. smithii* have undergone the evolution of life history specializations involving the polygenic systems controlling development time and diapause. Presumably, these patterns are the result of the sequential, post-pleistocene, adaptations of the species to northern climates and shorter seasons. The fast-developing, low-diapause genetic tendencies which can be isolated from northern populations are possibly retentions of older parts of the genetic systems adapted to subtropical climates. When hybrid populations formed from Ontario and Florida stocks are grown in simulated Ontario conditions of temperature and photoperiod, the Florida side of the cross exerts a dominance-like effect in the progeny. This dominance-like effect does not appear when the hybrids are reared in a simulated Florida environment.

When populations of *W. smithii* from places as distant as Ontario and Florida were crossed in the laboratory, no deleterious effects on the primary fitness characters were found. This result suggests that the diversifying, latitudinal specializations of life history and morphology known for this species are not concurrent with genetic evolution extensive enough, thus far, to promote the evolution of reproductive isolation and speciation.

Summary. The results of studies of the polygenic inheritance of development time and diapause in the pitcher-plant mosquito, *Wyeomyia smithii*, are reviewed. The data provide concrete examples of the genetic basis for a kind of life history variation that appears to be common in insects.

Crosses of latitudinally disparate stocks in the laboratory reveal that substantial genetic differences for life history patterns have evolved in *W. smithii*. This evolutionary divergence has not been so great as to produce deleterious effects on primary fitness characteristics in hybrid populations, and thus offers no indication that reproductive isolation and speciation is taking place. An attempt is made to reconstruct part of the process of the latitudinal diversification of the life history patterns among *W. smithii* populations during the post-pleistocene period.

Polygenic variation for development time and diapause combined with fluctuating-stabilizing selection appear to continually adapt *W. smithii* populations to an uncertain environment. An intermediate optimum in the fraction of diapause during the summer, and thus a mixed strategy or bet-hedging tactic, is favored. Data showing that the mixed strategy is present in the progeny of individual females throughout the summer is presented. A model of this adaptive process is developed. The model combines both ecological and genetic parameters.

Life history microevolution toward ever more asynchronous development of progeny may provide a generalized mode of resistance in response to human attempts to control pest insect populations.

Acknowledgments. I am grateful to Steve Orzack, S. D. Tuljapurkar, Tom Caraco, Ernst Caspari, John Moeur, William Etges and participants in the Evolution Seminar at the University of Rochester for discussions of the material in this paper. The studies reported here were supported by NSF Grant DEB-7724615.

References

Ayala, F. J.: Relative fitness of populations of *Drosophila serrata* and *Drosophila birchii.* Genetics 51, 527-544 (1965).

Bradshaw, W. E.: Photoperiodic timing of development in the pitcher-plant mosquito. Am. Zool. 11, 670-671 (1971).

Bradshaw, W. E.: Geography of photoperiodic response in a diapausing mosquito. Nature (London) 262, 384-386 (1976).

Bradshaw, W. E., Lounibos, L. P.: Photoperiodic control of development in the pitcher-plant mosquito, *Wyeomyia smithii.* Can. J. Zool. 50, 713-719 (1972).

Bradshaw, W. E., Lounibos, L. P.: Evolution of dormancy and its photoperiodic control in pitcher-plant mosquitoes. Evolution 31, 546-567 (1977).

Cohen, D.: A theoretical model for the optimal timing of diapause. Am. Nat. 104, 389-400 (1970).

Dingle, H., Brown, C. K., Hegmann, J. P.: The nature of genetic variance influencing photoperiodic diapause in a migrant insect. Am. Nat. 111, 1047-1059 (1977).

Fisher, R. A.: The Genetical Theory of Natural Selection. New York: Dover, 1958.

Hairston, N. G., Tinkle, D. W., Wilbur, H. M.: Natural selection and the parameters of population growth. J. Wildl. Manage. 34, 681-689 (1970).

Herzog, G. A., Philips, J. R.: Selection for a diapause strain of the bollworm, *Heriothis zea.* J. Hered. 67, 173-175 (1976).

Istock, C. A.: Fitness variation in a natural population. In: Evolution of Insect Migration and Diapause. Dingle, H. (ed.). New York: Springer-Verlag, 1978.

Istock, C. A.: The extent and consequences of heritable variation for fitness characters. In: Population Biology. Retrospect and Prospect. Oregon State University Colloquium: Biology. King, C. R., Dawson, P. S. (eds.). New York: Columbia Univ. Press, 1981 (in press).

Istock, C. A., Wasserman, S. S., Zimmer, H.: Ecology and evolution of the pitcher-plant mosquito. 1. Population dynamics and responses to food and population density. Evolution 29, 296-312 (1975).

Istock, C. A., Zisfein, J., Vavra, K.: Ecology and evolution of the pitcher-plant mosquito. 2. The substructure of fitness. Evolution 30, 535-547 (1976a).

Istock, C. A., Vavra, K., Zimmer, H.: Ecology and evolution of the pitcher-plant mosquito. 3. Resource tracking by a natural population. Evolution 30, 548-557 (1976b).

Lande, R.: The maintenance of genetic variability by mutation in a polygenic character with linked loci. Genet. Res. 26, 221-235 (1976).

Lang, J. T.: Relationship of fecundity to the nutritional quality of larval and adult diets of *Wyeomyia smithii*. Mosq. News 38, 396-403 (1978).

Lynch, C. B., Hoy, M. A.: Diapause in the gypsy moth: Environment specific mode of inheritance. Genet. Res. 32, 129-133 (1978).

Maynard Smith, J.: The effects of normalizing and disruptive selection on genes for recombination. Genet. Res. 33, 121-128 (1979).

Moeur, J. E., Istock, C. A. Ecology and evolution of the pitcher-plant mosquito. 4. Larval influence over adult reproductive performance and longevity. J. Anim. Ecol. (1980) (in press).

Morris, R. F.: Observed and simulated changes in genetic quality in natural populations of *Hyphantria cunea*. Can. Entomol. 103, 893-906 (1971).

Morris, R. F., Fulton, W. C.: Heritability of diapause intensity in *Hyphantria cunea* and related responses. Can. Entomol. 102, 927-938 (1970).

Oikarinen, A., Lumme, J.: Selection against photoperiodic reproductive diapause in *Drosophila littoralis*. Hereditas 90, 119-125 (1979).

Saul, S. H.: Fitness, adaptation and interdemic selection in populations of Drosophila. Ph.D. Thesis, University of Rochester, Rochester, N.Y., 1970.

Stearns, S. C.: Life-history tactics: a review of the ideas. Q. Rev. Biol. 51, 3-47 (1976).

Takahashi, F.: Generation carryover of a fraction of population members as an animal adaptation to unstable environmental conditions. Res. Popul. Ecol. 18, 235-242 (1977).

Thoday, J. M.: Components of fitness. Symp. Soc. Exp. Biol. 7, 96-113 (1953).

Thoday, J. M.: Natural selection and biological progress. In: A Century of Darwin. Barnett, S. A. (ed.). London: Heineman, 1958.

Wright, S.: Evolution and the Genetics of Populations, Vol. 1. Genetic and Biometric Foundations. Chicago, Ill.: Univ. Chicago Press, 1968.

Chapter 8

Alternative Life History Patterns in Risky Environments: An Example from Lacebugs

DOUGLAS W. TALLAMY and ROBERT F. DENNO

1 Introduction

In the 1960s and early 70s the evolution of life history patterns among organisms received a great deal of attention (reviewed by Stearns 1976). Theoretical explanations of life histories often treated observed patterns as products of single, causal systems. Three different though nonexclusive theories emerged: (1) the deterministic view based on the predictions of MacArthur and Wilson's (1967) r- and K-selection, (2) the bet-hedging hypothesis organized by Stearns (1976) from the stochastic models of Murphy (1968) and Schaffer (1974), and (3) the balanced-mortality hypothesis, so termed by Price (1974).

In the models of r- and K-selection, mortality and fecundity schedules do not fluctuate. K-selection is said to occur in resource-limited environments that favor the ability to compete and avoid predation, while r-selection occurs in environments favoring rapid population growth. Advocates (Cole 1954, Lewontin 1965, MacArthur and Wilson 1967, Gadgil and Bossert 1970, Pianka 1970, King and Anderson 1971) identify the density of a species with respect to its resources as the selective pressure capable of explaining two fundamental groupings of life history traits (see Table 8-1).

The bet-hedging hypothesis deals specifically with the consequences of fluctuating mortality schedules. When juvenile mortality fluctuates due to unpredictable environmental conditions, selection favors organisms with the same traits prediced under K-selection (Table 8-1). If fluctuations in juvenile mortality become more predictable because of stable conditions, organisms that can reproduce quickly but more than once are favored (Murphy 1968). In the case of unpredictable adult mortality (Schaffer 1974), selection in fluctuating and stable environments is predicted to parallel r- and K-selection respectively.

The balanced-mortality hypothesis explains life history evolution in terms of mortality suffered in a hostile environment. It holds that selection will balance high mortality levels by increasing egg production. Hence, if the probability that an organism will survive in a particular environment is low, selection will favor a compensatory increase in egg production. Cole (1954) suggests that the high fecundity of parasites

Table 8-1. The contrasting predictions of r- and K-selection and the bet-hedging hypothesis (as modified from Stearns 1976)

Stable environments (K-selection)	Fluctuating environments (r-selection)
Bet-hedging with adult mortality variable	
Delayed reproduction	Early age of first reproduction
Iteroparity	Semelparity
Small reproductive effort	Large reproductive effort
Few, large offspring	Many, small offspring
Long life	Short life
Bet-hedging with juvenile mortality variable	
Early age at first reproduction	Delayed reproduction
Iteroparity	Iteroparity
Large reproductive effort	Small reproductive effort
Short life	Long life
Large clutches	Small clutches
Few clutches	Many clutches

and marine organisms is an adaptation to a high probability of juvenile mortality. Similarly, when Price (1973, 1974) examined the parasite complex attacking the Swaine jack pine sawfly, he found that parasite fecundity was negatively correlated with the probability of survival of the host stage being attacked.

Bet-hedging, r- and K-selection, and the balanced-mortality hypotheses share a common assumption that has rarely been addressed in the literature (but see Wilbur et al. 1974, Stearns 1976). Each assumes that only particular combinations of life history traits coevolve under certain selection regimes. It has become increasingly obvious that theory organized on this assumption can not adequately explain the variation in life histories found in nature. A given package of life history traits is the product of many interacting selection pressures that maximize individual fitness through evolutionary responses to trade-offs among traits. There is no reason to expect that only one system of life history trade-offs is possible in particular environments. More likely, fitness is maximized in the same environment in a number of different ways.

In this report we discuss the life histories of four species of lacebugs as evidence that, contrary to the predictions of much of the historical literature, (1) life history traits evolve in response to many interacting selective pressures, and (2) more than one assemblage of traits may be adaptive under similar environmental conditions. The effects of predation and resource availability on the evolution of life histories are contrasted for lacebugs exploiting woodland hosts and lacebugs specializing on early successional plants.

2 Lacebug Biology

(1) *Corythucha ciliata* (Say), the sycamore lacebug, is restricted to *Platanus* sp. and occurs abundantly on *P. occidentalis* L. in the eastern United States. It overwinters as adults in the bark crevices of its host and, in College Park, Prince Georges County, Maryland, becomes active when the leaves are fully expanded in May. Small localized populations expand throughout the summer, spreading to adjacent uninhabited trees as the population increases. With five generations per summer, populations occasionally injure even large trees by September. Females wedge their eggs between large leaf veins and the leaf surface on the undersurface of leaves (Fig. 8-1). Clutches are small, each containing $\simeq$ 5 eggs. Young nymphs feed in aggregations of up to 50 individuals until maturity, but it is not uncommon for fourth and fifth instars to become solitary.

(2) *Corythucha pruni* Osborn and Drake, the cherry lacebug, exploits *Prunus serotina* Ehrh. in the wild but will also infest cultivated species of cherry (Bailey 1951). Clutches of $\approx$ 20 eggs are laid fully exposed on the undersurface of the leaf (Fig. 8-1) and hatch into gregarious nymphs. *C. pruni* overwinters as adults and has four generations per year in Maryland.

(3) *Corythucha marmorata* (Uhler), the chrysanthemum lacebug, is confined to early successional composites, feeding commonly on *Solidago, Aster* and *Ambrosia* (Bailey 1951). Only populations exploiting goldenrod (*Solidago*) were included in this study. Overwintering adults emerge in early spring and oviposit on the first available basal rosettes. Eggs are deeply embedded in the lower leaf surface along the midrib, often with several females contributing to the same egg mass (Fig. 8-1). Nymphs of all ages are gregarious. Up to five generations per year are common in Maryland.

(4) *Gargaphia solani* Heidemann, the eggplant lacebug, specializes in solanaceous plants, most commonly horsenettle (*Solanum carolinense* L.). It is a prolific species with large clutches and short generations, completing as many as eight generations a year in Virginia (Bailey 1951). Egg masses are cemented to the undersurface of leaves over a period of 3-4 days. As the mass is completed, females undergo a physiological suppression of egg production and begin to exhibit maternal behavior. Females remain with and aggressively defend their progeny throughout all five nymphal instars. After the young mature and disperse, females lay a second egg mass and resume their maternal behavior (Tallamy and Denno 1981).

3 Life History Patterns

The results of natural selection are expressed in differential survivorship and reproduction. Thus, the mortality suffered by herbivores on their particular host plants reflects in part the intensity of the selection pressures shaping herbivore life histories. We compared lacebug mortality schedules in natural stands of sycamore, cherry, goldenrod and horsenettle habitats by constructing survivorship curves for natural populations of the juvenile stages of each species. A minimum of 10 cohorts corresponding in initial size to the average size of the parent egg masses were monitored daily from egg to adult eclosion. To avoid the confounding effects of differential weather conditions, all four

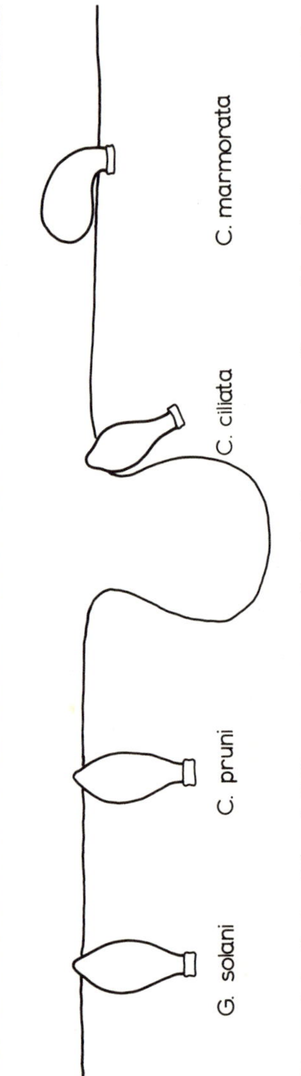

Figure 8-1. Ovipositional differences among four lacebug species. *Gargaphia solani*: egg cemented to leaf surface; *Corythucha pruni*: egg cemented to leaf surface; *C. ciliata*: egg wedged between large veins and leaf surface; *C. marmorata*: egg buried in leaf tissue except for protruding operculum.

species were studied simultaneously in College Park, Prince Georges County, Maryland, in 1978.

A Kruskal-Wallis ANOVA shows no significant difference in the percent survival at adult eclosion of *C. ciliata, C. pruni, C. marmorata* and *G. solani* (Fig. 8-2). Mortality rates are constant and statistically equal throughout juvenile development. Two explanations seem plausible: either the risks of exploiting all four hosts are equal, or each lacebug species has adapted to very different environmental risks with equal success. Evidence presented later in this paper supports the latter contention. The data must be viewed with caution, however. Selection coefficients of only 1% are sufficiently strong to cause detectable evolutionary changes in life history traits in 10 generations (Gill, pers. comm.), so ecologically imperceptible differences in survivorship may have considerable effects on the evolution of reproductive traits.

Before we can investigate further the evolution of life history patterns, the life history traits of the species studied must be quantified. Table 8-2 summarizes these traits as they were measured in controlled environmental growth chambers at 27 °C. Figure 8-3 presents the fecundity schedule of each lacebug through time. Heritability in survivorship-fecundity schedules and their determinants has been demonstrated in insects (Ohba 1967, Crovello and Hacker 1972, Dobzhansky et al. 1964), lizards (Tinkle 1967) and humans (Keyfitz 1968). These schedules are clearly responses to natural selection and reflect the evolutionary interaction between life histories and the environment.

The data suggest that, even though the survivorship schedules of the four lacebugs are similar, their life history traits have coevolved in very different ways. Several factors distinguish the life history of *G. solani* (on horsenettle) from the other species of lacebugs. It lays only two clutches, but each contains many small eggs. Total fecundity is low in *G. solani,* yet its intrinsic rate of increase is substantially higher than the remaining species. Perhaps the most important contribution to its high r is the early age of its first oviposition (Cole 1954, Lewontin 1965): at least five days earlier than that of the other species. Furthermore, *G. solani* is unique because its life history includes a comparatively advanced state of maternal behavior whereby females tend eggs and nymphs throughout their developmental period (Tallamy and Denno 1981).

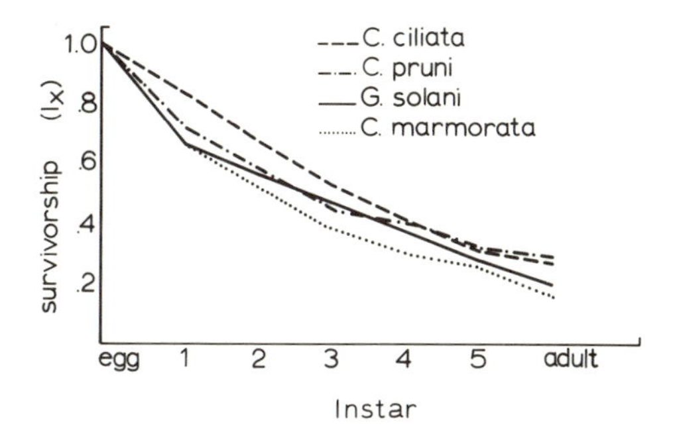

Figure 8-2. Survivorship schedules of juvenile *Corythucha ciliata* (sycamore lacebug), *C. pruni* (cherry lacebug), *Gargaphia solani* (eggplant lacebug), and *C. marmorata* (chrysanthemum lacebug).

Table 8-2. The assemblage of life history traits of four species of lacebugs reared at 27 °C. Note the general lack of conformity to the predictions of r- and K-selection, the balanced-mortality hypothesis and bet-hedging models

	G. solani	*C. marmorata*	*C. ciliata*	*C. pruni*
Host plant	horsenettle	goldenrod	sycamore	cherry
Parental care	present*,+	absent	absent	absent
Egg volume (mm^3/♀g)	10.5a	15.7*b	19.7+c	14.1b
Age at first oviposition (days)	20.3a	25.7*b	26.8bc	28.1c
Clutch size	68.4a	7.4*b	5.0b	11.7c
Number of clutches	1.8*a	18.1*b	33.2c	13.0d
Per capita fecundity	127.8*,°a	136.7*,°a	166.5*,°b	160.7*,°b
Intrinsic rate of increase (r-max)	0.171	0.119	0.129	0.124

*Traits not consistent with r- and K-predictions.
°Traits not consistent with the balanced-mortality hypothesis.
+Traits not consistent with the bet-hedging hypothesis.
a,b,c,d = means with different letters are significantly different (Duncan's Multiple Range; $P < 0.05$).

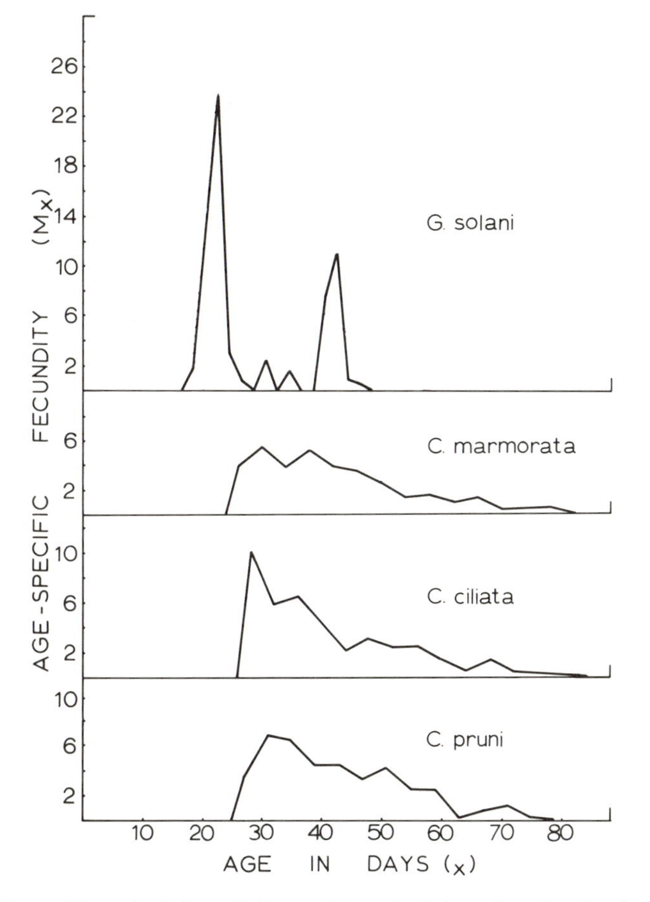

Figure 8-3. Fecundity schedules of *Gargaphia solani* (eggplant lacebug), *Corythucha marmorata* (chrysanthemum lacebug), *C. ciliata* (sycamore lacebug) and *C. pruni* (cherry lacebug) measured at 27 °C.

In contrast, *C. marmorata* (on goldenrod) lays several clutches over a much greater time period, matures more slowly and consequently increases at a lower rate. *C. ciliata* (on sycamore) exhibits the unlikely combination of many large eggs which are distributed in a large number of very small clutches. Though *C. pruni* (on cherry) is as fecund as *C. ciliata,* it lays small eggs in fewer but larger clutches. It is the slowest of the four species to mature. The intrinsic rate of increase of *C. marmorata* is similar to that of *C. ciliata* and *C. pruni,* but fecundity, clutch size and clutch number all differ significantly ($P < 0.05$).

4 A Test of General Theory: Predictions Based on r and K Models

The data presented in Table 8-2 provide a test of some of the predictions of r- and K-selection, bet-hedging, and the balanced-mortality hypothesis. The theory of r- and K-selection predicts that lacebugs occurring in early successional habitats are likely to be ravaged by density-independent mortality resulting from severe weather or generally unstable environmental conditions (Dobzhansky 1950, Cody 1966, Lack 1968). Accordingly, populations should be held below K, the carrying capacity of the environment, and respond to r-selection in order to compensate. An "r-strategy" is also predicted in species inhabiting old fields because of the relationship between the generation time of the insect (τ) and the length of time the host plant provides adequate resources (H) (Southwood et al. 1974). When τ/H approaches unity, as it may in insects that exploit ephemeral hosts, one generation has a minimal effect on the resources of the next generation. Thus, in the absence of selection against rapid population growth, r-related life histories are expected.

Converse arguments predict that lacebugs inhabiting forest hosts are not as exposed to density-independent mortality as those species exploiting weedy hosts. Instead, populations should respond to selection from density-dependent mortality factors such as competition and predation. Traits associated with a "K-strategy" are predicted. Also, since $\tau/H < 1$ in these species, selection should favor a life history that maintains population levels near K, but prevents overshooting that is detrimental to future generations.

A comparison of the assemblage of traits found in the four lacebugs with those commonly associated with r- or K-selected species (Pianka 1970) does not support the predictions. For example, *C. marmorata* has classic K-selected traits but thrives in an environment presumably reserved for r-strategists. *G. solani* achieves a rapid rate of increase in the predicted environment, but does so with parental care and low fecundity as integral parts of its life history. With the exception of high fecundity, the two forest species, *C. ciliata* and *C. pruni,* conform to the predictions of K-selection by maturing slowly and laying large eggs in many small clutches over a long period of time.

Supporters of r and K will readily point out the inadequacy of our test situation. We present no measure of population density with respect to resources and therefore do not directly test the theory. If, however, life histories evolve solely as a result of the closeness of population density to saturation, as r and K would lead us to believe, and if the association of traits predicted by r- and K-selection is indeed a general phenomenon, we would not expect to find the number of aberrations described above with or without density measurements.

There has been much support of r- and K-selection, both from field evidence correlating environmental patterns with reproductive patterns (Landahl and Root 1969, Pianka 1969, Tinkle et al. 1970, Bruce 1972, Force 1972, Abrahamson and Gadgil 1973) and from empirical studies (Solbrig 1971, Gadgil and Solbrig 1972, Solbrig and Simpson 1974, McNaughton 1975). Nevertheless, a growing body of data does not adhere to the predictions of r- and K-theory, suggesting that well-documented cases of such selection are the exception rather than the rule (Ramakrishnan 1960, Murphy

1968, Mertz 1971, 1975, King and Dawson 1972, Vepsäläinen 1974, 1978, Wilbur et al. 1974, Oka 1976). Surely a given set of environmental circumstances could select for the traits predicted by r and K. But, as our data indicate, similar environmental challenges may also give rise to a variety of other life history solutions.

4.1 Predictions Based on Balanced-Mortality Hypothesis

Unlike r- and K-theory, the balanced-mortality hypothesis disregards the distinction between density-independent mortality and density-dependent mortality, predicting that when the risks in exploiting a resource are high, fecundity will be correspondingly high and balance mortality levels (Skutch 1948, 1967, Cole 1954, Fretwell 1969, Wilson 1971, Price 1974). For example, according to the hypothesis, lacebugs that suffer high levels of mortality, whether from predators, host resistance or weather-related catastrophes, should have higher fecundities than species inhabiting environments with lower risks. Since exploiting sycamore or cherry habitats does not impart a greater probability of survivorship to lacebugs than exploiting horsenettle and goldenrod (Fig. 8-2), egg production should be equal in all four lacebugs.

The data clearly contradict these predictions (Table 8-2). *G. solani* and *C. marmorata* lay significantly fewer eggs than *C. ciliata* and *C. pruni*. We suggest that life history patterns cannot be predicted solely on the basis of mortality schedules because selection to decrease mortality can increase fitness equally as well as selection to balance mortality. The evolution of parental care, for instance, can effectively negate the risks of unusually harsh environments (Bro Larsen 1952, Highton and Savage 1961) or intense predation (Odhiambo 1960, Tallamy and Denno 1981). Similarly, adjustments in clutch size and frequency can reduce mortality from unpredictable environments (Vepsäläinen 1978) or predation (Stearns 1976).

4.2 Predications Based on the Bet-Hedging Hypothesis

The predictions of the bet-hedging hypothesis run contrary to the theory of r- and K-selection and support the arguments of Boer (1968), Cohen (1967) and Holgate (1967) on optimal clutch size. When environmental conditions make juvenile survival unpredictable, a single, large reproductive commitment, i.e., a "big bang strategy," may result in the total loss of offspring. It is more adaptive to increase the probability that at least some offspring survive by spreading reproductive effort over time and space. Such environments select for organisms that minimize the risk of leaving no progeny rather than maximize reproductive rates. Whether selection for variation in juvenile or adult survivorship has contributed to the life history patterns observed in the lacebugs is unknown. The data needed to test this hypothesis are difficult to obtain. Juvenile and adult mortality schedules must be replicated many times over several generations in order to permit meaningful statements about variation in the data. Nevertheless a liberal interpretation of Table 8-1 shows that the assemblage of traits predicted by bet-hedging is found in *C. pruni* and *C. marmorata,* provided juvenile mortality is variable in these species and goldenrod and cherry environments are unpredictable. We do not discount this possibility, but it is difficult to believe that similar selection has not occurred on *G. solani* and *C. ciliata.* The bet-hedging hypothesis should be credited with providing the most comprehensive set of predictions available on the evolution of life history patterns. That the hypothesis does not provide for alternative life history solutions to the described environmental conditions is discouraging.

5 Resource Availability

Differences observed in the life histories of the four lacebugs may (1) reflect different environmental risks associated with exploiting sycamore, cherry, goldenrod and horse nettle, (2) demonstrate different evolutionary solutions to the same environmental problems, or (3) represent a combination of both of these processes. The explanation requires an analysis of the environmental factors that influence reproductive success in these insects. One such factor, the availability of resources, is governed to a greater or lesser extent by the persistence of resources, i.e., the length of time resources remain suitable for feeding and development, resource dispersion, host resistance against its parasites and resource abundance. In this report we restrict the discussion to the evolutionary consequences of resource persistence and dispersion on life history patterns.

The accessibility of resources in time and space may be an important determinant of particular life history traits when viewed in terms of a trade-off between reproduction and dispersal. As resources become less persistent and/or more dispersed, organisms must devote increasing amounts of time and effort locating fresh resources. When resources are more persistent or clumped, however, dispersal mechanisms are less necessary, possibly leaving more time and effort for reproductive activities. The capacity for egg production in ichneumonid parasitoids attacking the Swaine jack pine sawfly decreases as the time and effort invested in locating suitable hosts increases (Price 1972). In addition, winged individuals in water striders polymorphic for wing length oviposit later, and thus increase at a slower rate than apterous individuals (Vepsäläinen 1978). In lacebugs, an evolutionary commitment to dispersal mechanisms such as wings, flight muscles, fat body stores or weight reduction could occur at the expense of reproduction.

Resource availability as measured by the persistence and dispersion of the host plant is difficult to quantify directly. Nevertheless, if it can be assumed that relative wing size reflects the flight demands on a species over evolutionary time, the degree of dispersal that is required to exploit a particular host and hence the availability of the host can be estimated. We generated a relative wing-loading index for each study species by dividing wing surface area of females by their body weight. Since the mesothoracic wings of lacebugs serve as protective structures whose actual contribution to flight is unknown, measurements of wing area were restricted to the metathoracic wings.

The result of wing-loading measurements (Table 8-3) are initially surprising. Mobility is commonly considered a characteristic of early successional species because of the relative ephemerality of their environments. Yet, according to the wing-loading index, *C. marmorata* and *G. solani* show a smaller commitment to flight than do *C. pruni* and *C. ciliata.* This provides an excellent example of how our perception of the envi-

Table 8-3. Wing-loading measurement for *Corythucha marmorata* (chrysantheum lacebug), *Gargaphia solani* (eggplant lacebug), *C. pruni* (cherry lacebug) and *C. ciliata* (sycamore lacebug)

C. marmorata	5.70a
G. solani	6.39a
C. pruni	7.43b
C. ciliata	7.94b

Metathoracic wing area (cm)/female weight (mg).
a,b = means (N = 10) with different letters are significantly different (Duncan's Multiple Range; $P<0.05$).

ronment is easily biased by viewing time and space on an inappropriate scale. Clearly the environment is defined by the organism. Insects perceive time and space in relation to their life span and size. Studies of insect environments must therefore be designed within this context.

Consider a group of host plants, spatially they are divided into leaves, small branches, large branches, single plants and entire plant populations. When a lacebug exhausts a leaf as a source of food or oviposition sites, it must move to an accessible leaf that can provide these resources. Interleaf distances are usually small and can easily be covered by walking. However, when all of the leaves on a branch tip are exhausted, the bug must move to an entirely different branch. On woody hosts interbranch distances may be so large that it becomes more efficient to fly than to walk.

Figure 8-4 presents a regression of lacebug metathoracic wing-loading, i.e., the evolutionary commitment to flight, against the effective distance between clusters of resources, be they complete horsenettle or goldenrod plants, or a cluster of cherry or sycamore leaves. It suggests that exploiting cherry and sycamore trees requires more flight than does exploiting goldenrod or horsenettle plants because of the spatial separation of the leaf resources. The life history of *C. ciliata* and *C. pruni* lends support to the argument. *C. ciliata* lays over 30 small clutches of eggs (Table 8-2) only along the major veins of a leaf. In addition to searches for food, this species must move considerably among its diffuse resources to locate limited oviposition sites. Goldenrod and horsenettle are both stoloniferous plants that characteristically grow in patches. Even when an entire plant is exhausted as a resource it is but a short, though presumably dangerous, walk to the next plant, and an even shorter, less risky flight. We suggest that one substantial difference in the allocation of time and effort in lacebugs is directly related to the spatial distribution of the resource. Energy allocated to flight is no longer available for reproduction.

6 Predation

From an evolutionary viewpoint the greatest difference between the life histories of these four lacebugs is the presence of maternal care in *G. solani.* Since parental care represents an extraordinary breakthrough in the adaptation of organisms to their environment, it is of great importance to understand the selective forces that maintain this trait. Predation has been cited as selecting for parental care if the risks to the parent do

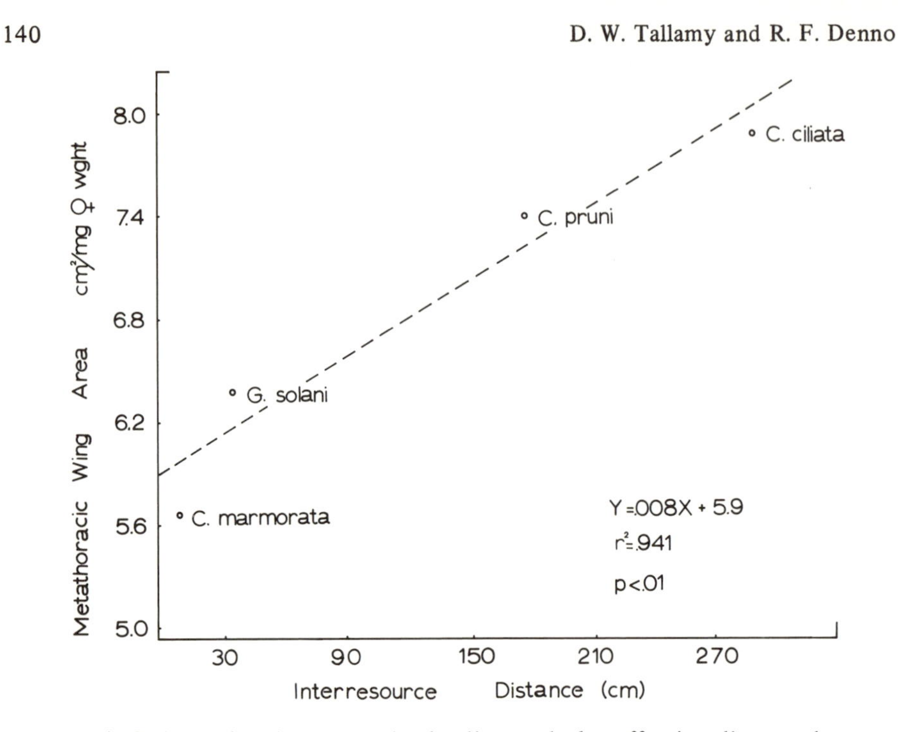

Figure 8-4. Relationship between wing-loading and the effective distance between available leaf resources in four species of lacebugs. *Corythucha marmorata* (chrysanthemum lacebug) and *Gargaphia solani* (eggplant lacebug) require only limited flight to exploit their weedy hosts. The large wings of *C. ciliata* (sycamore lacebug) and *C. pruni* (cherry lacebug) may be adaptations to dispersed resources.

not overrride the benefits to the young (Odhiambo 1960, Wilson 1975, Wood 1975, 1976). Tallamy and Denno (1981; submitted) support this hypothesis with strong evidence that, in spite of significant reductions in fecundity, maternal care in *G. solani* is adaptive in habitats with intense predation. Survivorship of juveniles with and without parent females was compared when predators were present and when predators were excluded from the system (Fig. 8-5). When predators were excluded there was no significant difference between the survivorship of nymphs attended by their mothers and orphaned nymphs. However, nymphs that were exposed to predation without the benefit of maternal protection suffered significantly higher mortality levels ($P < 0.05$) than nymphs guarded by their mothers. Consequently, in spite of a 50% reduction in fecundity (Tallamy and Denno, submitted), females that protected their young produced four times as many mature progeny as females that did not protect their young.

Because *C. marmorata, C. ciliata* and *C. pruni* do not care for their young, we must determine (1) whether predation is more intense in horsenettle habitats and/or (2) whether these species have evolved predator defenses other than maternal care. It is well known, through the work of Holling (1965, 1966, 1973), Hassell (1978) and others, that predators locate and remain in habitats that continually support the greatest number of prey. Since early successional fields are typically composed of large stands of a relatively few dominant plant species (Bard 1952), they should support

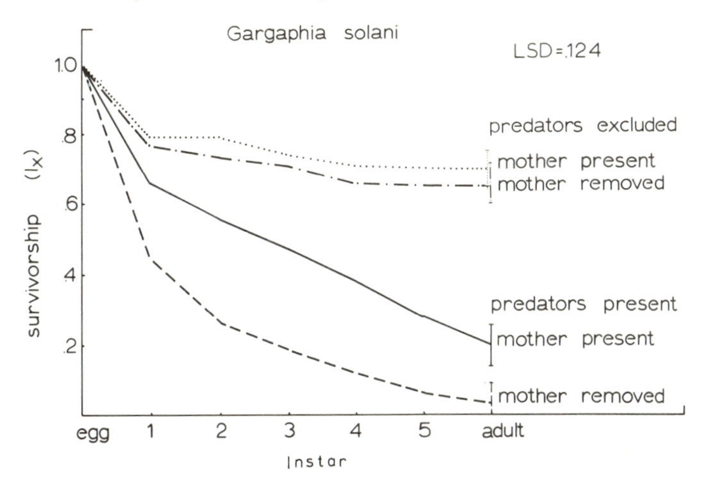

Figure 8-5. Effects of maternal care on the juvenile survivorship of the lacebug *Gargaphia solani* in the presence and absence of predators. LSD = Least Significant Difference multiple comparison test. Means differing by this interval or more are significantly different ($P < 0.05$).

herbivores and their predators in greater numbers than more diverse, late successional communities. Predation, therefore, should be more intense on *G. solani* and *C. marmorata* than on the species that exploit forest hosts.

To test this hypothesis, densities of lacebugs and known lacebug predators were recorded within randomly sampled sycamore, cherry, goldenrod and horsenettle leaves adjusted for area. The number of predators on each host was compared with an analysis covariance, treating lacebug as the covariant (Fig. 8-6). As predicted, there were significantly more predators in the early successional habitats than in the woodland habitats. Although predator power, i.e., the relative impact of the predator community

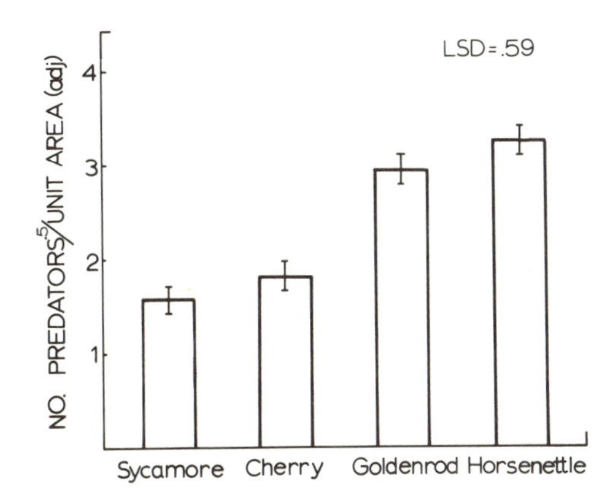

Figure 8-6. Differences in lacebug predator load (number of predators/lacebug) between sycamore, cherry, goldenrod and horsenettle habitats. LSD as in Figure 8-5.

on each lacebug species, was not assessed, the data support the hypothesis that maternal care in *G. solani* is an effective adaptation to the high predator densities inherent in horsenettle habitats. In sycamore and cherry habitats parental care may not be "cost-effective" due to lower predator densities.

Though predator densities are comparatively lower in sycamore and cherry habitats, it does not follow that predation does not influence life history evolution in the resident lacebugs. A review of Table 8-2 suggests that, in *C. ciliata* and to a lesser extent in *C. pruni,* selection has favored a diffuse pattern of oviposition whereby many small clutches are scattered over time and space. If predation were an important source of nymphal mortality, such a strategy should minimize the size of nymphal aggregations and thus the chances of entire aggregations being located and destroyed. Yet *C. ciliata* nymphs actively seek each other upon hatching and form aggregations far in excess of the average clutch size of five.

Predators of juvenile lacebugs can be divided into three groups: (1) obligate egg predators such as thrips, pirate bugs and mirids, (2) predators that include both eggs and nymphs in their diets, e.g., chrysopid larvae and coccinellids and (3) predators that feed only on nymphs, such as spiders, damsel bugs and assassin bugs. Table 8-4 presents the percentage of the predator community in each lacebug habitat that feeds exclusively on lacebug eggs or includes eggs in their diet. The predator community that exploits *C. ciliata* and *C. pruni* largely consists of egg predators. Thus, the ovipositional pattern observed in these lacebugs may increase the chances that clutches will not be discovered by predators and decrease the consequences to fitness in the event that a clutch is destroyed.

In addition to spreading its reproductive effort over time and space, *C. marmorata,* the goldenrod lacebug, apparently reduces the impact of egg predation by embedding its eggs in leaf tissue. However, since nymphs of this species face predator loads similar to those on *G. solani* without the benefits of parental care, it is possible that *C. marmorata* has evolved an alternative means of predator defense. This was tested by exposing third-instar nymphs of *G. solani* and *C. marmorata* in equal numbers to predators and recording the number of each species that was eaten after 12 hr. Two types of generalist predators were used, each representing a method whereby predatory insects consume their prey: (1) adult ladybird beetles (*Hippodamia* sp.) that mascerate prey with cusped mandibles and (2) larval lacewings (*Chrysopa* sp.) that capture, kill and suck out the contents of their prey with long, grooved mandibles. When the experi-

Table 8-4. The proportion of lacebug predators attacking *Gargaphia solani, Corythucha marmorata, C. pruni* and *C. ciliata* that feed exclusively on lacebug eggs or include eggs in their diet

	Proportion of predators that feed exclusively on eggs	Proportion of predators that include eggs in diet	Total proportion of predators that attack eggs
G. solani	7.7	39.8	47.5
C. marmorata	4.4	46.9	51.3
C. pruni	2.3	75.0	77.3
C. ciliata	20.4	68.3	87.7

ment was run using ladybird predators, *G. solani* nymphs were consumed significantly more often than *C. marmorata* nymphs ($P < 0.01$). However, when lacewing larvae were used as predators, both species were eaten in equal proportions.

The difference in prey selection between the two predator types is probably not related to differences in levels of sequestered chemical defenses that may exist between the lacebugs, but is more probably due to differences in lacebug behavior and/or external morphology. A close inspection of the external morphology of immature *C. marmorata* reveals an intricate complex of heavily sclerotized, three-pronged spines, each with a terminal barb, that is presented dorsally in five longitudinal rows extending the length of the body. Nymphs of this species are also dorsoventrally flattened. In contrast, *G. solani* nymphs are not flattened and have three longitudinal rows of single, non-barbed spines. Predators that chew their prey, such as ladybird beetles, may have more difficulty consuming a nymph that is essentially a flattened bed of barbed spines than a relatively unarmored, fat nymph. Predators, like lacewing larvae, that consume only the fluid contents of their prey are adept at inserting their prognathus mandibles between the cuticular spines of even the most well-protected nymphs.

The data suggest that intense predation has given rise to two quite different yet successful means of survival. *G. solani* protects its vulnerable eggs and nymphs from predators with effective maternal defense (Tallamy and Denno 1981), *C. marmorata* minimizes the effect of predation by burying its eggs in leaf tissue, spreading its reproductive effort over time and space and sporting protective nymphal armature. Predation that is less severe and directed toward the egg stage is countered by *C. ciliata* and *C. pruni* by dividing reproductive effort into many small clutches laid over an extended time period.

7 Summary and Conclusions

The problem of explaining the evolution of life histories is a formidable one. Past attempts to simplify the problem with general theory have more often than not hindered progress in this area. We have presented data on the reproductive biologies of four lacebug specialists that run contrary in parts to the theoretical predictions of r- and K-selection, bet-hedging and the balanced-mortality hypotheses. The discrepancies suggest that covarying life history traits evolve in responses to many interacting selective pressures, and that different combinations of traits may be equally adaptive in particular environments. Though we have attempted to explain several combinations of lacebug traits in terms of their evolutionary response to resource availability and predation, we by no means wish to imply that these are the only selective forces shaping life history patterns.

We have contrasted lacebug life history patterns as they have evolved under two sets of environmental circumstances: those encountered by herbivores exploiting woodland hosts, and those encountered by early successional herbivores. *C. ciliata* females, which produce large eggs attractive to egg predators, reduce predation on offspring by laying small clutches of eggs as they move from leaf to leaf throughout their adult lives. The dispersal of eggs over space and time and the exploitation of dispersed leaf resources within sycamore hosts may lead to proximate and ultimate trade-offs between dispersal and reproduction in this species. *C. pruni* uses resources that present similar environmental challenges. This species is as fecund as *C. ciliata* but lays smaller eggs in fewer, larger clutches.

G. solani and *C. marmorata* overcome environmental risks associated with early successional habitats in remarkably different ways. Since the resources of horsenettle and goldenrod are accessible without large commitments to dispersal, at least in the short term, the trade-off between dispersal and reproduction in *G. solani* and *C. marmorata* may be minimized. *G. solani* lays most of its eggs early in life, at one time and place, and channels its remaining efforts into parental protection against predators. Maternal care is an expensive trait because its induction seriously reduces fecundity (Tallamy and Denno, submitted). Therefore, to maximize fecundity, selection should minimize the time invested in maternal care. Since nymphs require maternal protection throughout their development, selection has apparently minimized the amount of time females guard each brood by increasing the developmental rate of juveniles. Also, since it requires no more time to guard a large group of nymphs than a small group (Tallamy and Denno, submitted), large clutches have been favored in this species. By contrast, *C. marmorata* survives similar predation pressures because it spreads its reproductive effort over time and space, buries its eggs in leaf tissue and is protected by complex spines against hypognathus predators during the vulnerable nymphal stages.

With these results in mind, it is difficult to justify the emphasis on predicting "the optimal" assemblage of reproductive traits that should evolve under certain conditions, when it appears that several different combinations of life history traits can be adaptive under similar environmental circumstances.

Acknowledgments. John Davidson, David Inouye, William Mellors, Douglass Miller and the late Robert Baker criticized earlier drafts of this report. Susan Smith and Alexandra Tallamy typed the manuscript. To these people we are appreciative.

Scientific Article No. A2889, Contribution No. 5943, of the Maryland Agricultural Experiment Station, Department of Entomology.

References

Abrahamson, W. G., Gadgil, M. D.: Growth form and reproductive effort in goldenrods (*Solidago*, Compositae). Am. Nat. 107, 651-661 (1973).

Bailey, N. S.: The Tingoidea of New England and their biology. Entomol. Am. 31, 1-140 (1951).

Bard, G. E.: Secondary succession on the piedmont of New Jersey. Ecol. Monogr. 22, 195-215 (1952).

Bro Larsen, E.: On subsocial beetles from the salt-marsh, their care of progeny and adaptation to salt and tide. Trans. Ninth Int. Congr. Entomol., Amsterdam 1951, 1, 502-506 (1952).

Bruce, R. C.: Variation in the life cycle of the salamander *Gryinophilus porphyriticus*. Herpetologica 28, 230-245 (1972).

Cody, M. L.: A general theory of clutch size. Evolution 20, 174-184 (1966).

Cohen, D.: Optimizing reproduction in a randomly varying environment when a correlation may exist between the conditions at the time a choice has to be made and the subsequent outcome. J. Theoret. Biol. 16, 1-14 (1967).

Cole, L. C.: The population consequences of life history phenomena. Q. Rev. Biol. 29, 103-137 (1954).

Crovello, T. J., Hacker, C. S.: Evolutionary strategies in life table characteristics among feral and urban strains of *Aedes aegypti* (L.). Evolution 26, 185-196 (1972).

den Boer, P.: Spreading of risk and stabilization of animal numbers. Acta Biotheor. 18, 165-194 (1968).

Dobzhansky, T.: Evolution in the tropics. Am. Sci. 38, 209-221 (1950).

Dobzhansky, T., Lewontin, R., Pavolovsky, O.: The capacity for increase in chromosomally polymorphic and monomorphic populations of *Drosophila pseudobscura*. Heredity 19, 597-614 (1964).

Force, D. C.: r- and K-strategies in endemic host-parasitoid communities. Bull. Entomol. Soc. Am. 18, 135-137 (1972).

Fretwell, S. D.: The adjustment of birth rate to mortality in birds. Ibis 3, 624-627 (1969).

Gadgil, M., Bossert, W. H.: Life historical consequences of natural selection. Am. Nat. 104, 1-24 (1970).

Gadgil, M., Solbrig, O. T.: The concept of r- and K-selection: evidence from wild flowers and some theoretical considerations. Am. Nat. 106, 14-31 (1972).

Hassell, M. P.: The Dynamics of Arthropod Predator-Prey Systems. Princeton, N.J.: Princeton Univ. Press, 1978.

Highton, R., Savage, T.: Functions of the brooding behavior in the female red-backed salamander, *Plethodon cinereus.* Copeia 1, 95-98 (1961).

Holling, C. S.: The functional response of predators to prey density and its role in mimicry and population regulation. Mem. Entomol. Soc. Can. 45, 3-60 (1965).

Holling, C. S.: The functional response of invertebrate predators to prey density. Mem. Entomol. Soc. Can. 48, 1-86 (1966).

Holling C. S.: Resilience and stability of ecological systems. Ann. Rev. Ecol. Syst. 4, 1-24 (1973).

Holgate, P.: Population survival and life history phenomena. J. Theor. Biol. 14, 1-10 (1967).

Keyfitz, N.: Introduction to the Mathematics of Population. Reading: Addison-Wesley, 1968.

King, C. E., Anderson, W. W.: Age specific selection. II. The interaction between r- and K- during population growth. Am. Nat. 105, 137-156 (1971).

King, C. E., Dawson, P. S.: Population biology and the *Tribolium* model. In: Evolutionary Biology, Vol. 5. Dobzhansky, Hecht, Steere (eds.). New York: Academic Press, 1972.

Lack, D.: Ecological Adaptations for Breeding in Birds. London: Methuen, 1968.

Landahl, J. T., Root, R. B.: Differences in the life tables of tropical and temperature milkweed bugs, genus *Oncopeltus* (Hemiptera: Lygaeidae). Ecology 50, 734-737 (1969).

Lewontin, R. C.: Selection for colonizing ability. In: The Genetics of Colonizing Species. Baker, Stebbins (eds.). New York: Academic Press, 1965.

MacArthur, R. H., Wilson, E. O.: The Theory of Island Biogeography. Princeton, N.J.: Princeton Univ. Press, 1967.

McNaughton, S. J.: r- and K-selection in *Typha.* Am. Nat. 109, 251-261 (1975).

Mertz, D. B.: Life history phenomena in increasing and decreasing populations. In: Statistical Ecology, Vol. 2. Patil, Pielou, Waters (eds.). University Park, Pa.: Penns. State Univ. Press, 1971, pp. 361-392.

Mertz, D. B.: Senescent decline in flour beetle strains selected for early adult fitness. Physiol. Zool. 48, 1-23 (1975).

Murphy, G. I.: Pattern in life history and the environment. Am. Nat. 102, 391-403 (1968).

Odhiambo, T. R.: Parental care in bugs and non-social insects. New Sci. 8, 449-451 (1960).

Oka, H. I.: Mortality and adaptive mechanisms of *Oryza perennis* strains. Evolution 30, 380-392 (1976).

Ohba, S.: Chromosomal polymorphism and capacity for increase under near optimal conditions. Heredity 22, 169-185 (1967).

Pianka, E. R.: Sympatry of desert lizards (*Ctenotus*) in Western Australia. Ecology 50, 1012-1030 (1969).

Pianka, E. R.: On r- and K-selection. Am. Nat. 100, 592-597 (1970).

Price, P. W.: Parasitoids utilizing the same host: Adaptive nature of differences in size and form Ecology 53, 190-195 (1972).

Price, P. W.: Reproductive strategies in parasitoid wasps. Am. Nat. 107, 684-693 (1973).

Price, P. W.: Strategies for egg production. Evolution 28, 76-84 (1974).

Ramakrishnan, P. S.: Ecology of *Echinochloa colunum* Linn. Proc. Ind. Acad. Sci. 52, 73-99 (1960).

Schaffer, W. M.: Selection for optimal life histories: the effects of age structure. Ecology 55, 291-303 (1974).

Skutch, A. F.: Do tropical birds rear as many young as they can nourish? Ibis 91, 430-455 (1949).

Skutch, A. F.: Adaptive limitation of the reproductive rate of birds. Ibis 109, 579-599 (1967).

Solbrig, O. T.: The population biology of dandelions. Am. Sci. 59, 686-694 (1971).

Solbrig, O. T., Simpson, B. B.: Components of regulation of a population of dandelions in Michigan. J. Ecol. 63, 473-486 (1974).

Southwood, R. T. E., May, R. M., Conway, G. R.: Ecological strategies and population parameters. Am. Nat. 108, 791-804 (1974).

Stearns, S. C.: Life history tactics: a review of the ideas. Q. Rev. Biol. 51, 3-47 (1976).

Tallamy, D. W., Denno, R. F.: Maternal care in *Gargaphia solani* (Hemiptera: Tingidae). Anim. Behav. (in press). 1981.

Tallamy, D. W., Denno, D. F.: Life history trade-offs in *Gargaphia solani* (Hemiptera: Tingidae): the cost of reproduction. Ecology (submitted).

Tinkle, D. W.: The life and demography of the side-blotched lizard, *Uta stansburiana*. Misc. Publ. Mus. Zool. Univ. Mich. 132, 1-182 (1967).

Tinkle, D. W., Wilbur, H. M., Tilley, S. G.: Evolutionary strategies in lizard reproduction. Evolution 24, 55-74 (1970).

Vepsäläinen, K.: The life cycles and wing lengths of Finnish *Gerris* Fabr. species (Heteroptera: Gerridae). Acta Zool. Fenn. 141, 4-73 (1974).

Vepsäläinen, K.: Wing dimorphism and diapause in *Gerris:* Determination and adaptive significance. In: The Evolution of Insect Migration and Diapause. Dingle, H. (ed.). New York: Springer-Verlag, 1978.

Wilbur, H. M., Tinkle, D. W., Collins, J. P.: Environmental certainty, trophic level, and resource availability in life history evolution. Am. Nat. 108, 805-817 (1974).

Wilson, E. O.: The Insect Societies. Cambridge, Mass.: Belknap Press, 1971.

Wilson, E. O.: Sociobiology The New Synthesis. Cambridge, Mass.: Belknap Press, 1975.

Wood, T. K.: Defense in two pre-social membracids (Homoptera: Membracidae). Can. Entomol. 107, 1227-1231 (1975).

Wood, T. K.: Alarm behaviour of brooding female *Umbonia crassicornis* (Homoptera: Membracidae). Ann. Entomol. Soc. Am. 69, 340-344 (1976).

PART THREE

Life Histories and Nonequilibrium Populations

Much of the literature on community structure has implicated interspecific interactions, especially competition, as an organizing force. A subsequent extension of this kind of thinking has frequently led ecologists to view the evolution of life history traits in an arena characterized by limited resources, stable populations, intense species interactions and density-dependent mortality. While this is apparently the case for some systems, for others equilibrium conditions are often too hastily assumed on the basis of insufficient data. The chapters in this part discuss several communities of herbivorous insects whose species fluctuate or vary in time and space in ways that are somewhat inconsistent with equilibrium or competition hypotheses.

In Chapter 9, R. F. Denno, M. J. Raupp and D.W. Tallamy examine the organization and associated life histories of a temperate guild of leafhoppers and planthoppers on Salt Meadow Hay (*Spartina*), a grass that grows abundantly in tidal marshes along the eastern seabord of North America. They discuss niche diversification and the density relationships of species, provide information on sources of mortality, and elaborate on the frequency of catastrophic events that selectively reduce or eliminate certain species from the guild. Furthermore, they associate the species' susceptibility to catastrophe and ability to colonize and persist in grass patches with particular combinations of life history traits. Lastly, Denno et al. discuss latitudinal changes in the equilibrium status, species richness, and life history trait make-up of the assemblage of sap-feeding insects with regard to an intermediate disturbance hypothesis. By contrast, in Chapter 10, D. R. Strong questions the role that competition plays, in the Neotropics, in structuring hispine beetle communities on *Heliconia* plants, and discusses other factors that influence their life histories. Beetle density, limitation of plant resources, nutritional and defensive properties of plants, agonistic and displacement behavior, and parasites are considered possible factors in the shaping of life histories and community organization of these beetles.

Complexity and patchiness per se in resources like host plants may result in nonequilibrium conditions. New patches are created by a diversity of processes, while others become extinct. Herbivores that colonize first are faced with a very different set of constraints from those arriving later to face a complex of already established herbivores and predators. As Denno et al. show, certain species may be selectively eliminated from patches while others remain by virtue of their life histories. As Whitham (Chapter 1) and Edmunds and Alstad (Chapter 2) suggest, variability in host plant quality in itself presents major problems for herbivorous insects. Emerging from this discussion is a dynamic picture of resources in both space and time, one where the notion of equilibrium populations and communities seems at the very least questionable. In the chapters that follow, the spatial problems posed by host plants as islands are considered for colonizing herbivores, as are life history solutions to those problems and resulting species distributions.

In Chapter 11, D. Simberloff experimentally addresses the question of what combination of life history characteristics confer good colonizing ability by studying the assemblage of insects associated with red mangrove islands in the Florida Keys. He examines the relationship between the ability to reach and the ability to persist on isolated islands, and finds a correlation between the two for a large group of species that occur often there. Although frequently cited as a mechanism influencing species abundances and distributions, competition on islands remains for the most part unproven, and biogeographic data implicating competition as an organizing force is to date inconclusive. Factors determining successful colonization and persistence thus remain largely unknown, and their elucidation presents a challenge for future "laborious autecological research," as Simberloff indicates. Also, as with the lacebugs studied by Tallamy and Denno, there may be different life history solutions to the problems of colonization and persistence on scattered resources.

In Chapter 12, R. P. Seifert also considers biogeographic issues, especially those which arise from a consideration of host plants as ecological "islands." He examines milkweed insects in temperate and tropical areas and notes that, while some island effects such as species-area relationships occur, distance effects do not. He concludes that milkweeds do act as ecological islands for their associated insects, but that ecological island patterns in general are different from true island patterns. However, the effects of host plant phenology, structure, defense and taxonomy on ensuing species-area relationships or island patterns in general are poorly known. Furthermore, the constraints these host plant factors place on the life histories of associated insects, subsequent species interactions and resulting community patterns will be revealed only by a continuing rigorous, autecological approach.

Chapter 9

Organization of a Guild of Sap-feeding Insects: Equilibrium vs. Nonequilibrium Coexistence

ROBERT F. DENNO, MICHAEL J. RAUPP and DOUGLAS W. TALLAMY

1 Introduction

The great majority of ecological literature assumes that communities are in equilibrium and that interspecific competition plays a primary role in their organization (e.g., Lotka 1931, Gause 1934, Slobodkin 1961, MacArthur and Wilson 1967, Schoener 1974, Hutchinson 1977). For insects, several hypotheses have been invoked when attempting to explain coexistence under equilibrium conditions. McClure and Price (1975) suggest that coexistence in a guild of sycamore-feeding leafhoppers may be partially explained by frequency-dependent competitive ability, since at high densities each species adversely affects its own fitness more than that of its competitors. However, by far the most common explanation for equilibrium existence is resource partitioning, whereby species reduce interspecific competition by exploiting different aspects of resources (e.g., Connell 1961, MacArthur 1972, Cody 1974, Schoener 1974, Harper 1977, Diamond 1978). As Strong points out in his contribution (Chapter 10), competition has been invoked primarily by ecologists working with vertebrates, plants or marine organisms rather than insects.

By contrast, Hutchinson (1953) suggested two situations where species may coexist without achieving a stable equilibrium. First, nonequilbrium coexistence can be attained by organisms with several generations per year, where changing environmental conditions shift species-dominance and prevent any one species from excluding the other. Secondly, Hutchinson (1953) as well as Skellam (1951) emphasize that a nonequilibrium community is possible if catastrophic events destroy existing habitats while new ones are simultaneously created. Under these circumstances, competitively weaker species (those that would be excluded under equilibrium conditions) may coexist because of their increased ability to disperse and colonize vacant sites. We would add to this category any species that by virtue of any particular life history trait(s) (dispersal or otherwise) is favored during catastrophy.

Recently, the notion of equilibrium and competition mediated communities has been challenged (see Dayton 1971, Caswell 1978). For instance, Connell (1978) provides evidence that tropical rain forests and coral reefs, though traditionally regarded as highly ordered equilibrium communities, are actually seldom near equilibrium because of either frequent disturbances or more gradual climatic changes. Thus, intermediate levels of disturbance prevent the elimination of inferior competitors. The findings of Hubbell (1979) on the organization of tropical dry forests are also conssistent with a nonequilibrium hypothesis.

There is considerable literature on niche differentitation in coexisting parasites of fish, leading one to the conclusions that competition is keen and species-packing tight in these communities (Holmes 1973, Hair and Holmes 1975, Price 1980 and many references therein). Yet, Kennedy (1977) concluded that many fish parasites exist in nonequilibrium conditions and that interspecific competition plays little role in regulating their populations.

Strong (1977a, 1977b; Chapter 10) has studied rolled-leaf hispine beetles that feed in the New World tropics on the leaves of the monicot, *Heliconia.* Food and space within rolled leaves were well in excess of the levels that would adversely affect the beetles, and interspecific competition was not shown to structure communities. Furthermore, the most recent and convincing hypothesis to account for the coexistence of eight species of leafhoppers on sycamore trees in Illinois invokes nonequilibrium conditions (McClure and Price 1976, Price 1980). In a recent volume on parasites (broadly defined as an organism living in or on another living organism—and including most conventional parasites and parsitoids as well as phytophagous organisms), Price (1980) concludes that there is little evidence to suggest that competition has been an organizing force in parasite communities, that nonequilbrium conditions often prevail, and that the partitioning of resources and subsequent specializations we see result from evolutionary constraints other than competition.

In this report we hope to elucidate some of the factors that are important in organizing phytophagous insect communities by examining a guild of sap-feeding insects (mostly leafhoppers and planthoppers) that feed on the intertidal marsh grass, *Spartina patens* Lois. Several characteristics of this system make it attractive for examining community processes. The grass is widespread and is a dominant component of salt marsh vegetation from New Hampshire to Florida. It often occurs in pure stands that exist as an archipelago of "islands" varying in size from several square meters to huge (> 10-ha) patches set in a "sea" of other marsh plants. The sap-feeding insects on *S. patens* are abundant, diverse (≃ 10 species) and for the most part host-specific. Also, several very different life history patterns are represented in the guild.

Here, we review evidence suggesting that the community of sap-feeders associated with *S. patens* is occasionally in equilibrium and is organized by competitive interactions. Then we provide additional evidence suggesting that the community is not in equilibrium at least during some years and that coexistence is achieved by the selective elimination of certain species during catastrophic events creating vacant sites for others to exploit.

2 Distribution, Structure, and Growth Dynamics of *Spartina patens*

North of North Carolina, in the mid-Atlantic and New England states, the high intertidal marsh association (mean high water level to $\simeq$ 0.5 m above) is quite simple, where it is covered primarily by the grass, *Spartina patens* (Ait.) Muhl. (Miller and Egler 1950, Blum 1968, Redfield 1972). The high marsh is inundated only by the highest spring tides and storm tides. In this geographic area, variation in the elevational relief of the marsh surface results in an archipelago of usually pure *S. patens* patches ranging in size from just a few square meters to mammoth "islands" (> 3 ha) surrounded by another grass species, *Spartina alterniflora* Lois, that dominates the low intertidal marsh. In the Carolinas and south to Florida and along the Gulf Coast, *S. patens* occurs as a broken fringe of vegetation along the high marsh (Mobberley 1956, Adams 1963). Here, stands of grass can be pure, but rarely attain sizes greater than 1000 m^2. Also, in this portion of its range, plants tend to be a bit larger and more robust.

S. patens is a slender-culmed grass with narrow, convoluted blades (Mobberley 1956, Blum 1968). The living culms of *S. patens* grow and project through a thick (5-20-cm), dead horizon of prostrate culms and blades resulting from the previous year's growth. New culms, shaped like vertical awls, first protrude through the thatch in spring. As the season progresses, older leaf blades separate from newer, upright ones by bending at the sheath-blade junction. As subsequent blades fold back in this fashion, they make contact with the surface of the dead thatch. By summer, the dead thatch becomes overlaid with an entanglement of living leaf blades. Further prostration occurs in summer and fall, when the culms of *S. patens* fold over at a weak area in the stem, which coincides approximately with that portion of the stem that is included in and surrounded by the dead thatch (Blum 1968). Prostration usually occurs in a mosaic fashion, leaving behind small patches of somewhat erect culms. Flowering occurs during summer and early fall.

If the structure of *S. patens* is examined during summer, one finds an uppermost layer of living, partially prostrate grass overlying a dead horizon of dry culms from the previous year. Beneath this dry horizon is a layer of entangled moist culms and blades two and three years old. Between the moist layer of culms and the marsh surface is a horizon of decaying grass older than three years. Often, near the base of individual plants, the lower layer of thatch fails to contact the surface of the marsh because it is suspended by the dense crowns of the individual plants. Young tillers commonly protrude from the crown into the open space. Also, foraging voles (*Microtus*) create a labyrinth of tunnels underneath the thatch and appear to play an important role in maintaining this open space at certain locations. See Fig. 9-3 for a stylized representation of the grass system.

Maximum live biomass of *S. patens* occurs between mid-July and mid-August on mid-Atlantic state marshes, with no apparent relationship between grass patch size and either maximum live biomass or date of peak biomass (Fig. 9-1A). However, large patches of grass (those > 1001 m^2) do possess significantly greater amounts of dead thatch than small (< 1000 m^2) patches (F-test, $P < 0.01$) (Fig. 9-1B). The difference is particularly large early in the season in June prior to decay.

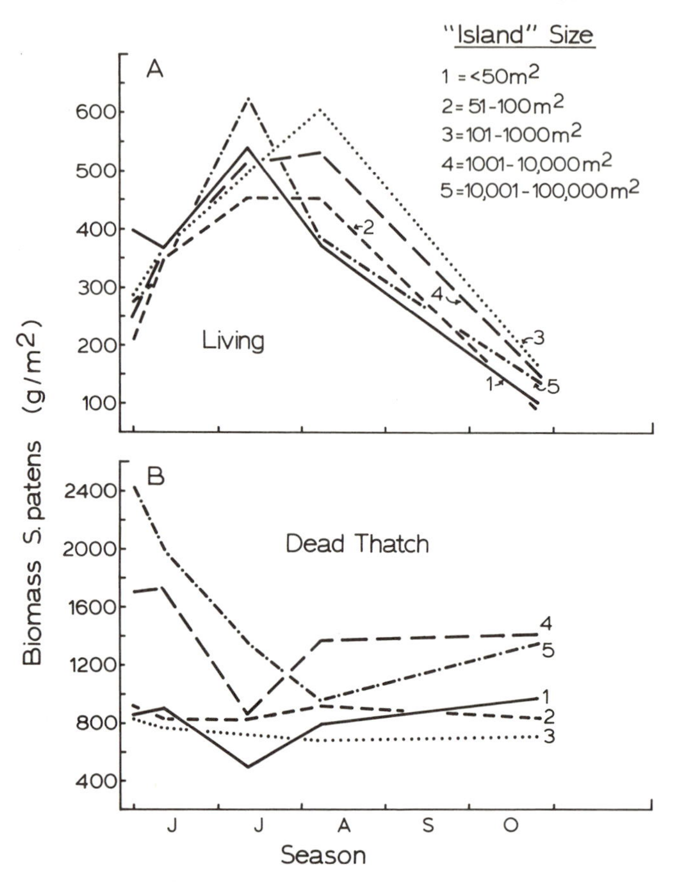

Figure 9-1. Seasonal changes in living biomass (A) and dead thatch (B) in 5 different "island" (stand) sizes of *S. patens* at Tuckerton, New Jersey, during 1978. Means are based on 3 samples in at least 3 "islands" for each size category except the largest (5) that consisted of only a single large island. For details on grass biomass sampling see Denno 1977.

We used percent crude protein (determined by standard macro-Kjeldahl analysis) as an index of grass quality. Percent crude protein was highest during May (12-14%), after which it dropped rapidly until mid-July, whereafter it remained continuously low (Fig. 9-2). Grass patches of all sizes showed this pattern, and there was no apparent relationship between patch size and quality.

3 Guild of Sap-feeders on *Spartina patens*

Although *S. patens* is fed upon by a number of functional feeding groups of insects (e.g., strip-feeders and stem borers), by far the most abundant and diverse herbivorous insects are sap-feeders in the suborder Auchenorrhyncha (leafhoppers and planthoppers) (Davis and Gray 1966, Denno 1976, 1977, 1980). Seven resident sap-feeding insects occur abundantly and develop on *S. patens* on New England and mid-Atlantic

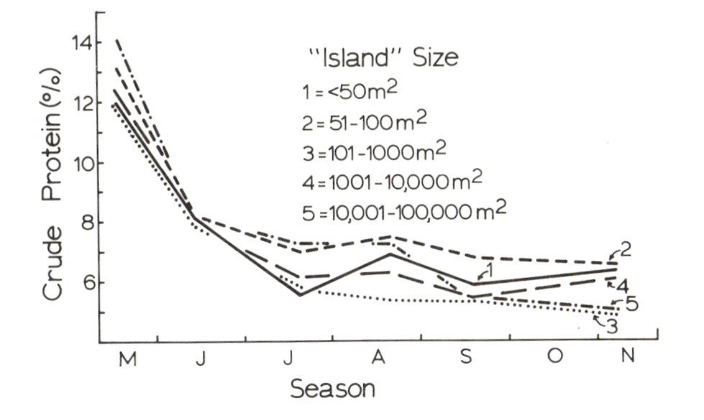

Figure 9-2. Seasonal changes in the quality (% crude protein) in 5 different "island" (stand) sizes of *S. patens* at Tuckerton, New Jersey, during 1979. Sampling scheme as in Fig. 9-1. The % crude protein in each sample was determined using the standard macro-Kjeldahl analysis (see Horwitz 1965).

state marshes south through Virginia. Of these, four are planthoppers (Delphacidae), *Delphacodes detecta* (Van Duzee), *Neomegamelanus dorsalis* (Metcalf), *Tumidagena minuta* McDermott, and *Megamelus lobatus* Beamer, and two are leafhoppers (Cicadellidae), *Amplicephalus simplex* (Van Duzee) and *Destria bisignata* (Sanders and DeLong). There is also the fulgoroid (Issidae), *Aphelonema simplex* (Uhler). The leafhopper, *Hecalus lineatus* (Uhler), occurs rarely on mid-Atlantic marshes, but is abundant only at the extreme northern end of the range of *S. patens* in New Hampshire.

Aphelonema decorata (Van Duzee) replaces *A. simplex* on *S. patens* in South Carolina south through Florida. *Tumidagena terminalis* (Metcalf) replaces *T. minuta* on North Carolina marshes, and is common on *S. patens* south through Florida. These congeneric pairs contain extremely similar taxa (both morphologically and ecologically) that may in fact represent the ends of step clines. *D. detecta, N. dorsalis, M. lobatus* and *D. bisignata* all range south through Florida. Several other sap-feeders are residents on *S. patens,* but because they are rare they are not included in this discussion (see Denno 1977).

4 Ecotope Differentiation in the Guild of Sap-feeders

In this section we review the resource-use patterns and body-size relationships of the common sap-feeders associated with *S. patens* on mid-Atlantic marshes. Resource partitioning was studied using one habitat factor (marsh elevation) and two niche factors (microhabitat distribution within the grass system and temporal utilization) (details on insect sampling are available in Denno 1980).

To determine the microhabitat distribution of the various sap-feeders, the grass system was divided into five rather distinct vertical zones (see the definition of zones in the stylized representation of the grass system at the top of Fig. 9-3): Zone 5 consisted of the seed heads and terminal blades of the upright living grass. Zone 4 comprised subterminal blades and stems of the living grass, and was like zone 5 except that most plant parts were shaded. Prostrate living culms and blades knocked over by winds formed zone 3. A thatch layer of horizontal dead culms and blades through which passes the basal portion of ensheathed, vertical, living culms defined zone 2. Zone 1 consisted of the crowns and tillers of the grass beneath the thatch layer.

Four sap-feeders, *D. detecta, N. dorsalis, A. simplex,* and *Am. simplex* occurred primarily in the upper stratum (zones 3-5) of the grass system above the thatch layer (Fig. 9-3). A second group of three species, *T. minuta, M. lobatus* and *D. bisignata,* inhabited primarily the lower stratum of the grass, in and beneath the thatch layer (zones 1 and 2) (Fig. 9-4).

Sap-feeders of the upper stratum had very similar seasonal distributions (Fig. 9-5A-D). For *D. detecta, N. dorsalis* and *A. simplex,* populations began increasing in May, peaked in September, and then decreased rapidly. Only *Am. simplex* peaked earlier in the season during July. *T. minuta, M. lobatus* and *D. bisignata,* the lower-stratum residents, also showed very similar seasonal distributions, but as a group tended to peak

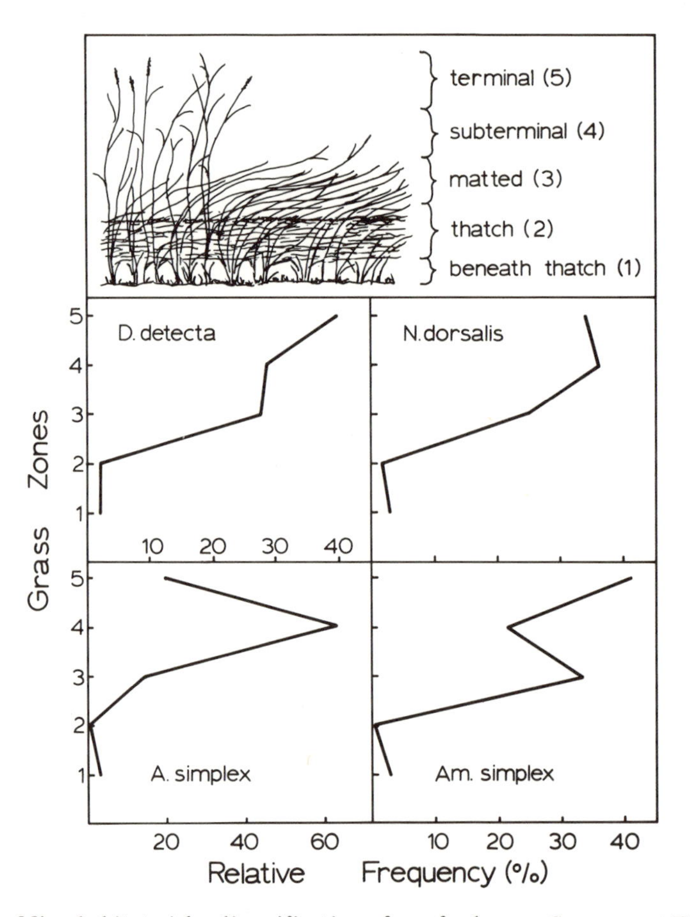

Figure 9-3. Microhabitat niche diversification of sap-feeders on *S. patens* at Tuckerton, New Jersey. Relative frequency of *D. detecta, N. dorsalis, A. simplex* and *Am. simplex* in terminal (5), subterminal (4), matted (3), thatch (2) and beneath thatch (1) zones of *S. patens* (see Denno 1980 for details on sampling).

about one month later than upper-stratum species (Fig. 9-5E-G). Populations increased in June, peaked during October or November and then declined rapidly. For details on the phenology of these species see Denno (1977, 1980) and Raupp and Denno (1979). The similarity in phenology of all of the sap-feeders is probably the result of selective presssures associated with peak biomass and quality of the grass (see Figs. 9-1A, 9-2) (Denno 1980).

Denno (1980) also determined the distribution of the sap-feeders along an elevational gradient in similar-sized patches of *S. patens* that occurred from approximately mean high water level (MHW) to 25 cm above. All sap-feeders but *N. dorsalis* were most abundant on patches of grass that occurred within 10 cm of MHW (Fig. 9-6). Only *N. dorsalis* predominated on, and was for the most part restricted to, patches at the upper end of the elevational range of the grass (Fig. 9-6).

Hutchinson (1959) measured the feeding apparatus of closely related species when they were sympatric and allopatric. He found that the ratio of the largest to the smallest dimension was ≅ 1.0 when the species were allopatric, but where they co-occurred the species differed by a factor of 1.2-1.4. He tentatively concluded that a difference of about 1.2-1.4 was necessary for two congeners to coexist. Some work tends to support this conclusion. For example, Rosenzweig and Sterner (1970) and Brown (1975) suggested that coexistence in heteromyid rodent communities is permitted by body size (mass) differentials. Similarly, various sizes of sap-feeders may be able to negotiate the microhabitat in different ways, and oviposition, feeding and hiding sites that are available to a small species may not be to a larger one. Consequently, sap-feeders that differ in size may use resources differently and, thereby, reduce competition. We used

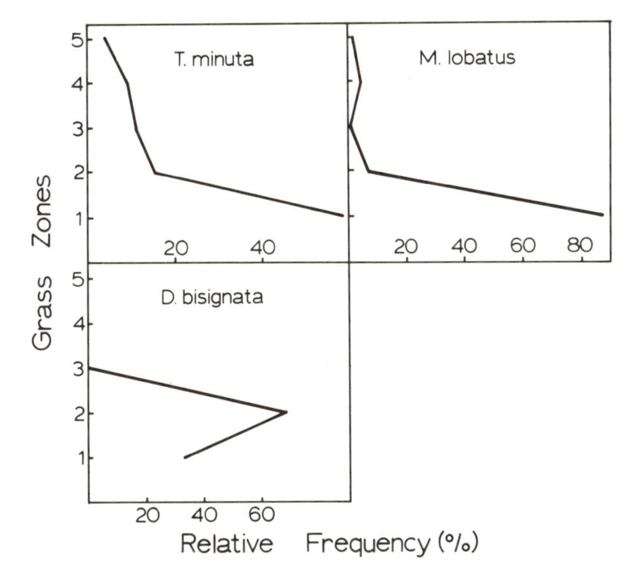

Figure 9-4. Microhabitat niche diversification of sap-feeders on *S. patens* at Tuckerton, New Jersey. Relative frequency of *T. minuta, M. lobatus* and *D. bisignata* in terminal (5), subterminal (4), matted (3), thatch (2) and beneath thatch (1) zones of *S. patens* (see Denno 1980 for details on sampling).

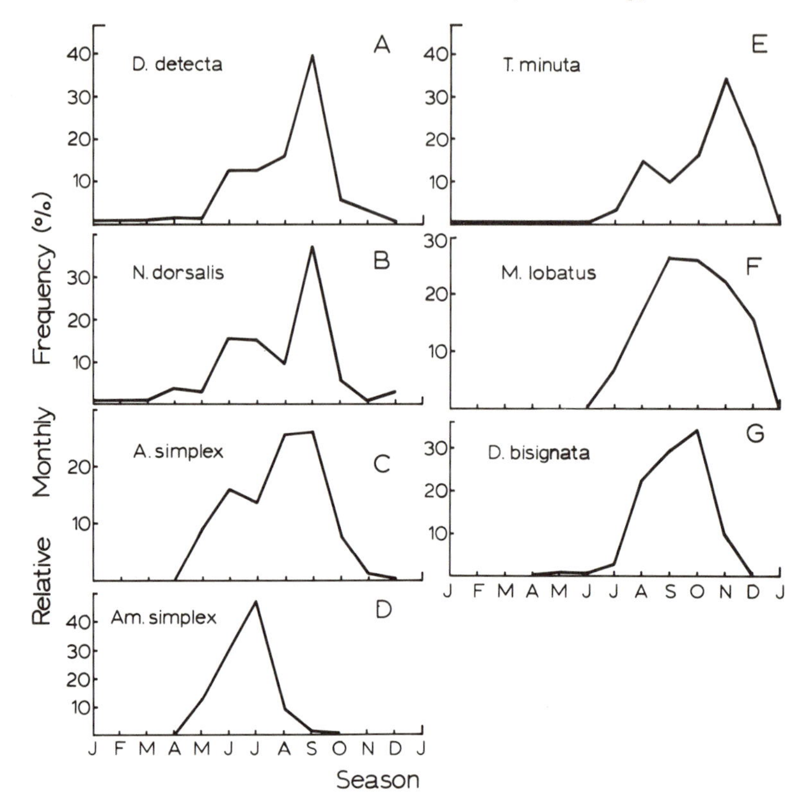

Figure 9-5. Seasonal niche diversification at Tuckerton, New Jersey. Relative monthly frequency of sap-feeders on *S. patens* (see Denno 1980 for details on sampling).

a body size ratio of 1.3 (midpoint of Hutchinson's 1.2-1.4 range) as a measure of the difference necessary to permit coexistence. Because stylets can be retracted through the beak and coiled within the head of these animals, beak length is not an appropriate measure of potentially important differences in the way sap-feeders use resources.

The mean dry adult body mass of the various *S. patens* sap-feeders ranged from 0.11 to 0.78 mg (Table 9-1). When the entire assemblage was considered as a unit, three species pairs (*M. lobatus* and *D. detecta, D. detecta* and *D. bisignata,* and *A. simplex* and *Am. simplex*) failed to differ in body weight by a factor of at least 1.3 (Ratio A of Table 9-1). However, when species were grouped by microhabitat into upper- and lower-zone residents (Ratios B and C, resp., of Table 9-1), only *A. simplex* and *Am. simplex* differed by a ratio (1.15) of less than 1.3.

Based on MacArthur's (1972) $d > \sqrt{\sigma_1^2 + \sigma_2^2}$ analysis that estimates the probability for competitive exclusion along resource dimensions (microhabitat, season and elevation), and Hutchinson's (1959) ratio of $\simeq 1.3$ as an indication of the amount of morphological (body size, in this case) difference necessary to permit coexistence, all species but two (*Am. simplex* and *A. simplex*) differ enough in the way they use *S. patens* to allow for co-occurrence (Fig. 9-7) (see Denno 1980). The sap-feeders segregate well into two subguilds (upper- and lower-strata species) along the microhabitat

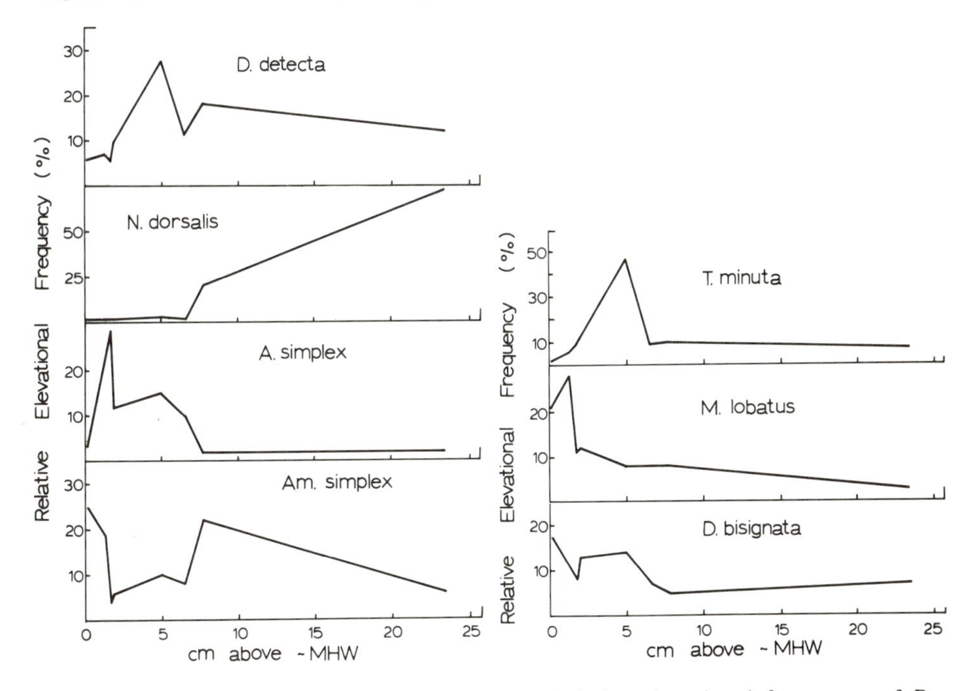

Figure 9-6. Elevational habitat diversification. Relative elevational frequency of *D. detecta, N. dorsalis, A. simplex, Am. simplex, T. minuta, M. lobatus* and *D. bisignata* on patches of *S. patens* along a gradient from ≃ mean high water level (MHW) to 25 cm above at Tuckerton, New Jersey (see Denno 1980 for elevation determination and sampling).

Table 9-1. Adult dry body mass (mg) and body mass ratios[a] (large species/small species) for the resident sap-feeders on *S. patens*

Species	Dry weight ($\bar{x} \pm \sigma$)	Ratio A	Ratio B	Ratio C
N. dorsalis	0.11 ± 0.008	1.36		
T. minuta	0.15 ± 0.018	1.67	2.73	1.67
M. lobatus	0.25 ± 0.032	1.20		1.36
D. detecta	0.30 ± 0.037	1.13		
D. bisignata	0.34 ± 0.012	2.00	2.27	
A. simplex	0.68 ± 0.036	1.15	1.15	
Am. simplex	0.78 ± 0.044			

[a]NOTE: Ratios between all species (A), upper-stratum species (B) and lower-stratum species (C).

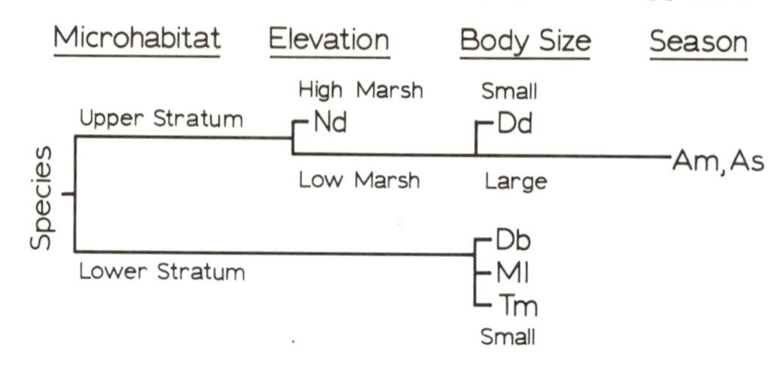

Figure 9-7. Niche differentiation in the guild of *S. patens* sap-feeders. Branching under microhabitat and elevational dimensions indicates that there is sufficient resource partitioning ($d > \sqrt{\sigma_1^2 + \sigma_2^2}$) to permit coexistence. Branching under the body-size category indicates that sap-feeders differ sufficiently in body size (ratio of large to small species > 1.3) to coexist. Nd, *Neomegamelanus dorsalis;* Dd, *Delphacodes detecta*; Am, *Amplicephalus simplex*; As, *Aphelonema simplex*; Db, *Destria bisignata*; Ml, *Megamelus lobatus*; Tm, *Tumidagena minuta* (see Denno 1980).

dimension. Of the lower-stratum species none are separated by the elevational dimension, but all differ in body size by a ratio > 1.3, implying that coexistence is possible. *N. dorsalis* displaces the other upper-stratum species on the elevational dimension. The body size of *D. detecta* is sufficiently small to separate it from *Am. simplex* and *A. simplex*. However, *A. simplex* and *Am. simplex* are not sufficiently different from one another along any niche or habitat dimensions, nor do they differ in body size by a factor > 1.3.

Even with the inherent weaknesses of these analyses in mind (see Simberloff and Boecklen 1981), we feel we have shown that *Am. simplex* and *A. simplex* share very similar niches, and the question remains as to how they both persist. Their position seems even more precarious when one considers that their smaller instars are very similar in size to the larger nymphs and adults of *D. detecta,* creating a situation where competition is potentially further intensified. Also, Denno (1980) found significant negative relationships between the densities of *A. simplex, Am. simplex* and *D. detecta,* the upper-stratum residents (Fig. 9-8A and B). The regressions were generated on the basis of samples taken in May 1976, when the nymphs of *A. simplex* and *Am. simplex* occur with the adults of *D. detecta* and when body sizes of the three species are similar. It is curious that both *A. simplex* and *Am. simplex* maintain such low densities relative to those of *D. detecta,* and that their densities increase only at the very lowest densities of *D. detecta.* This suggests that interference competition rather than resource depletion may govern the system. Also, there is a negative relationship between the densities of *A. simplex* and *Am. simplex,* two very similar-sized species that commonly co-occur during July (Fig. 9-8C). This finding takes on added significance because there is a positive relationship between the densities of the upper- (*D. detecta*) and lower-strata (*T. minuta*) dominants in the same (July) samples (Fig. 9-8D). Because of their fidelity to their respective stratum, these two species rarely encounter one another. Possibly this positive density relationship results from a common response to variation in host plant quality.

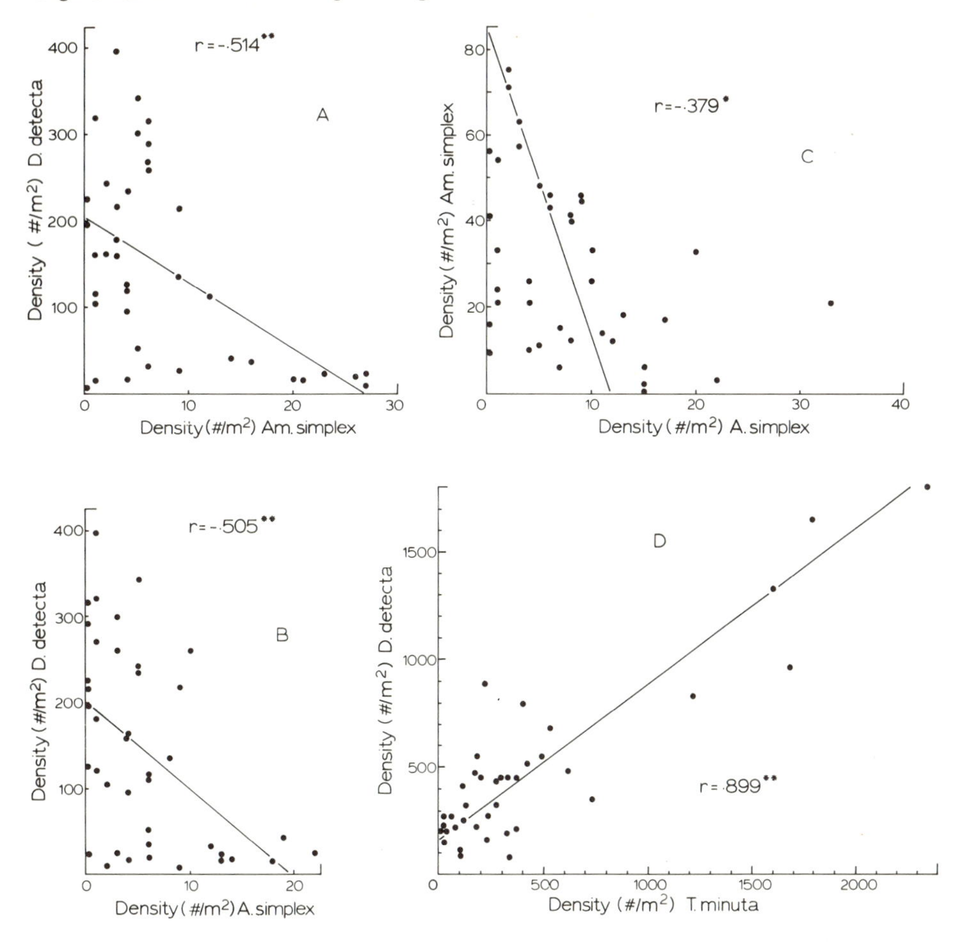

Figure 9-8. Relationship between the density of *D. detecta* and the densities of *Am. simplex* (A) and *A. simplex* (B) in samples taken during May 1976, *Am. simplex* and *A. simplex* (C) and *D. detecta* and *T. minuta* (D) in samples taken during July 1976, in stands of *S. patens* at Tuckerton, New Jersey (* $P < 0.05$; ** $P < 0.01$).

When the density relationships of *Am. simplex* and *D. detecta* are viewed on a different time scale, similar results are obtained. For instance, there is a significant negative relationship between the mean annual densities of these two species measured during five seasons in large patches of grass in New Jersey (Fig. 9-9). Thus, the data presented so far on resource partitioning, body sizes and density relationships are all consistent with the argument that this community of sap-feeders is organized by competitive interactions, and appears to be overpacked (see Denno 1980 for additional supportive data). By looking at the life history characteristics, patch-size exploitation patterns and long-term within-patch population fluctuations of these sap-feeders that we provide below, we hope to answer questions concerning the apparent overpacked guild, the equilibrium status of the community, and the importance of the resource-partitioning data presented so far.

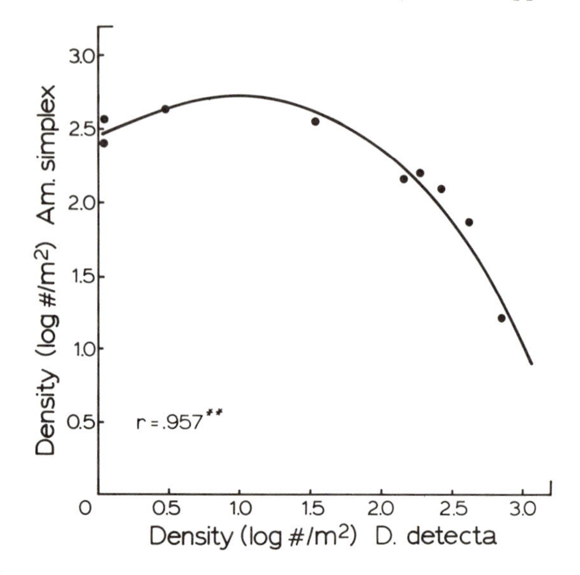

Figure 9-9. Relationship between the mean spring densities (May, June) of *Am. simplex* and *D. detecta* for a 6-year period (1974-1979) at Tuckerton, New Jersey. Two large (> 1000 m^2) stands of *S. patens* were sampled each year except during 1974 (** $P < 0.01$).

5 Dispersal Abilities, Overwintering Styles and Oviposition Characteristics of Sap-feeders

Dispersal ability and other life history characteristics discussed below affect a species' ability to track resources in time and space, and ultimately reflect on the potential for achieving equilibrium in populations. Wing dimorphism is a common phenomenon in planthoppers (Delphacidae and Issidae) and some leafhoppers (Cicadellidae), and can be used to elucidate dispersal ability. There are short-winged, flightless individuals (brachypters) and individuals with fully developed wings (macropters) that fly. Because of their ability to escape deteriorating resources and colonize new ones (see Denno and Grissell 1979, Denno et al. 1980), macropters should be adaptive in unstable habitats where resources fluctuate greatly and better alternatives are available elsewhere. On the other hand, the brachypters of several species of planthoppers have been shown to have a higher fecundity and oviposit at an earlier age than macropters (Tsai et al. 1964, Kisimoto 1965, Nasu 1969, May 1971, Mochida 1973). As a result, the reproductive potential of brachypters is greater than for macropters. Also, because of their flightlessness, brachypters are more likely to remain on the immediate resource. For these reasons, brachypters can more effectively exploit persistent resources (Denno 1976, 1978, 1979, Denno and Grissell 1979, Denno et al. 1980), particularly those that occur in climatically harsh habitats, where it is adaptive to balance high mortality with increased fecundity. Consequently, by examining the proportion of wing forms in a population, the potential for colonization can be estimated for a species.

Wing form is mediated by an individual's response during the nymphal stage to various proximate environmental cues (e.g., host plant quality and crowding) that measure the probability of current resource deterioration (Kisimoto 1956, 1965, Mochida 1973, Denno 1976). When the fitness of an individual can be increased by moving from deteriorating to more favorable resources, selection has apparently favored a low threshold, at which a developmental switch responds to environmental cues resulting in the production of a long-winged individual. By contrast, where conditions are not better "elsewhere" than "here," selection favors a high-threshold response resulting in a short-winged animal (see Denno 1976, Denno and Grissell 1979, Denno et al. 1980).

Historically, salt marshes and most of their included vegetation types (see Denno and Grissell 1979, Denno et al. 1980 for one exception) have been considered persistent systems (Southwood 1962, Brinkhurst 1963, Johnson 1969), and brachypters are by far the most abundant wing-morph in planthopper populations (Denno 1976, 1978). The wing-morph frequencies of most of the planthoppers and leafhoppers in the *S. patens* guild of sap-feeders are consistent with this general pattern. Populations of all species but one, *Am. simplex,* are composed of at least 86% brachypters (Table 9-2). Consequently, the dispersal capabilities for most individuals of most species are greatly reduced. Only *Am. simplex* completely retains the ability to fly.

In New Jersey, the planthoppers *D. detecta* and *N. dorsalis* are trivoltine and overwinter as active nymphs on the marsh surface or in the dead thatch of *S. patens* (Denno 1977). Although they do gain some protection within the thatch during winter, they are still exposed to very cold temperatures, and even ice, in harsh seasons. *T. minuta* has an identical life history, except that it completes only two generations per year. *D. bisignata* and *M. lobatus* also complete two generations per year, but overwinter as eggs that are inserted deep into the culms or blades of the grass with a saw-like ovipositor (see Fig. 9-10). *Am. simplex* (Fig. 9-10) and *H. lineatus* also insert their eggs into the grass where they remain hidden during winter, but these two species are univoltine. Although *A. simplex* overwinters as eggs, it lacks the saw-like ovipositor possessed by the other species and is unable to insert its eggs (Fig. 9-10). Eggs remain exposed on the vegetation. With regard to overwintering styles, the sap-feeders on *S. patens* fit into two categories: There are species that overwinter either as exposed eggs or nymphs (*D. detecta, N. dorsalis, T. minuta* and *A. simplex*), and should be vulnerable to harsh winter conditions. *D. bisignata, Am. simplex, H. lineatus* and *M. lobatus* insert their eggs, and should be less susceptible to the rigors of winter. The overwintering styles and generational information of these sap-feeders are summarized in Table 9-3.

6 Density Patterns of Sap-feeders along a Size Gradient of *S. patens* Patches

Patchiness per se may promote nonequilibrium conditions. For instance, resource area has been shown to have major effects on immigration and extinction rates and subsequently on the density of individuals and number of species that are able to obtain in a particular patch (MacArthur and Wilson 1967, Simberloff 1969, 1976, Simberloff and Wilson 1969, 1970, Raupp and Denno 1979, Rey 1981). Raupp and Denno (1979) suggest that, in general, extinction rates of sap-feeders are higher on small compared to

Table 9-2. Flightlessness in populations of sap-feeders

Species	% Brachypters
Am. simplex	0
D. detecta	86
N. dorsalis	88
M. lobatus	97
T. minuta	>99
A. simplex	>99
D. bisignata	100 (♀)
H. lineatus	100

large "islands" of *S. patens* because mortality associated with tidal flooding is greater there. Distance of resource from source area has also been proposed to have an effect on immigration rate (MacArthur and Wilson 1967). However, Rey (1981), working with the community of insects (also dominated by sap-feeders) associated with *Spartina alterniflora* in Florida, was unable to find a statistically significant effect of isolation on "island" immigration rate. We have concentrated on area rather than distance effects in our study by sampling sap-feeders on ≃ 25 "islands" of *S. patens* ranging in size from $< 50\ m^2$ to patches $\simeq 100{,}000\ m^2$. All "islands" sampled were located within 25 m of large ($> 1000\ m^2$) source "islands" that were included in a large archipelago of patches on a marsh in Tuckerton, New Jersey.

Differences in the life history characteristics (e.g., migration ability and overwintering style) of the *S. patens* sap-feeders probably influence each species' ability to colonize and remain on small compared to large patches of grass. Denno and Grissell (1979) and Denno et al. (1980) show that the brachypterous morph in dimorphic populations of planthoppers is virtually immobile and unable to move more than a few meters. By contrast, macropters are capable of long-distance flight, and are able to colonize distant, more favorable patches. Also Denno (1977), by defaunating large plots of *S. patens* and measuring recolonization, has shown that fully winged species like *Am. simplex* are excellent colonists. Thus, species that produce mostly flightless brachypters

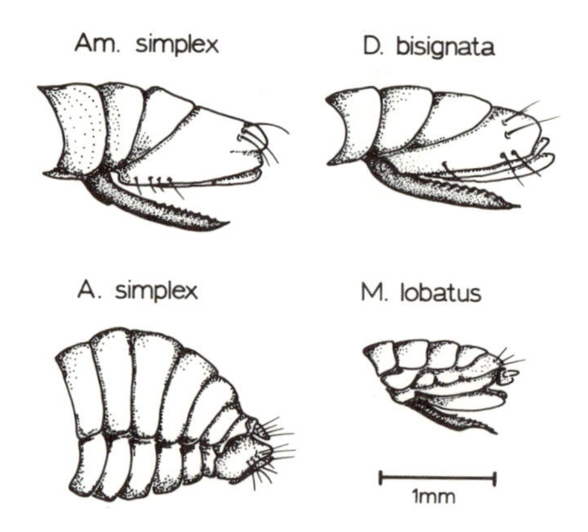

Figure 9-10. Abdomens of *Am. simplex, D. bisignata* and *M. lobatus* with saw-like ovipositors capable of inserting eggs into vegetation. *A. simplex* lacks a piercing ovipositor and is unable to embed its eggs.

Table 9-3. Generations and overwintering stages of sap-feeders

Species	#Gen/Yr	Overwintering Stage	Site and exposure of overwintering stage
T. minuta	2	nymph	marsh surface (exposed)
D. detecta	3	nymph	marsh surface (exposed)
N. dorsalis	3	nymph	marsh surface (exposed)
A. simplex	2	egg	on vegetation (exposed)
D. bisignata	2	egg	in vegetation (embedded)
M. lobatus	2	egg	in vegetation (embedded)
Am. simplex	1	egg	in vegetation (embedded)
H. lineatus	1	egg	in vegetation (embedded)

(see Table 9-2) should be poor colonists (have low immigration rates) and be under-represented on small, isolated "islands."

Overwintering style should also affect the ability of sap-feeders to remain on small islands. Species that overwinter as exposed eggs or nymphs should incur more winter mortality than those that overwinter as eggs embedded in vegetation (see Table 9-3). Extinction rates should be particularly high on small islands for exposed species, because there is less protective thatch there during the winter season (see Fig. 9-1B). Tallamy and Denno (1979) show that the overwintering success of sap-feeders with exposed stages (see *Tumidagena minuta*) is comparably low in plots of grass where the protective thatch has been removed.

Based on life history data, several predictions can be made concerning the success of sap-feeders on small as opposed to large patches of *S. patens*. Species with low immigration rates (those that are mostly brachypterous) and high extinction rates (in part those that overwinter in an exposed stage) should be least likely to maintain populations on small islands. Sap-feeders with these characteristics are *T. minuta* and *A. simplex*. Populations of both are composed of > 99% brachypters and both overwinter as exposed stages (see Tables 9-2 and 9-3). The responses of these two species are as predicted; both become increasingly more abundant as grass patch size increases (Fig. 9-11A and B). The differential effects of winter mortality on small "islands" are particularly evident for *T. minuta* (Fig. 9-11A). Population size was very small on small "islands" (< 100 m^2) during May prior to reproduction. By September, population size had increased on small "islands," but remained small compared to that attained on large "islands." The fact that populations do build rapidly on small "islands" suggests either that reproductive increase is high for *T. minuta* or that colonization is occurring. The former is probably more likely because most adults are brachypterous, although nymphs may be dispersed to small "islands" during tidal inundation. Nymphs have a waxy substance on their cuticle that allows them to float on the water's surface. Nevertheless, high extinction rate during winter is probably the major factor dictating the poor success of *T. minuta* on small islands.

A. simplex showed the same trend of abundance along the island size gradient, but never occurred on the smallest islands of grass (< 50 m^2) and achieved high densities only on the very largest of grass patches (Fig. 9-11B). Denno (1977) documents the poor colonizing ability of this sap-feeder. Thus, low immigration rate and high extinction rate during winter act in concert on small islands dictating its poor success there.

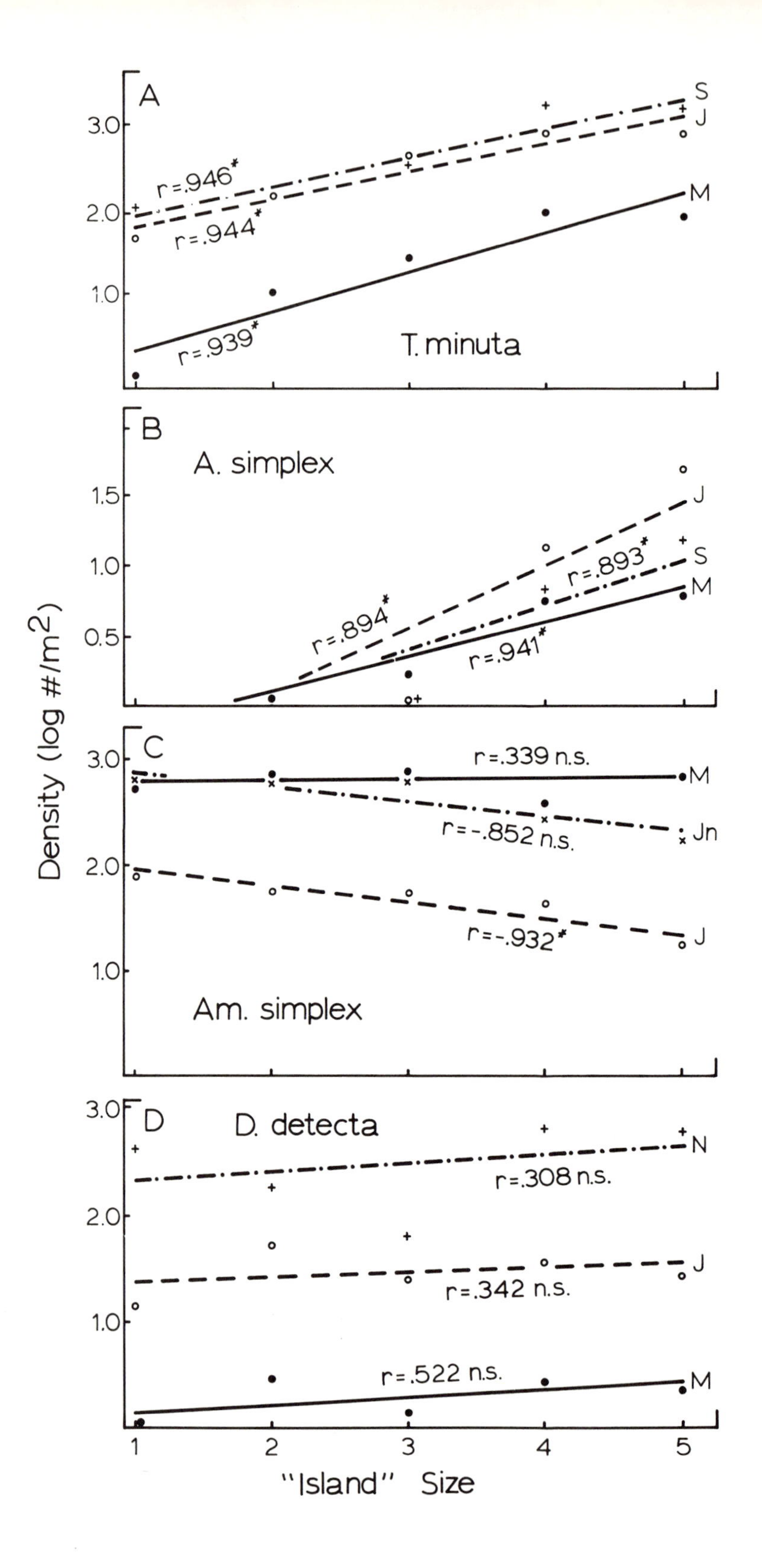

A
S
J
M
r=.946*
r=.944*
r=.939*
T. minuta
B
A. simplex
J
S
M
r=.894*
r=.893*
r=.941*
C
r=.339 n.s.
M
Jn
r=-.852 n.s.
J
r=-.932*
Am. simplex
D
D. detecta
N
r=.308 n.s.
J
r=.342 n.s.
M
r=.522 n.s.
Density (log #/m2)
"Island" Size

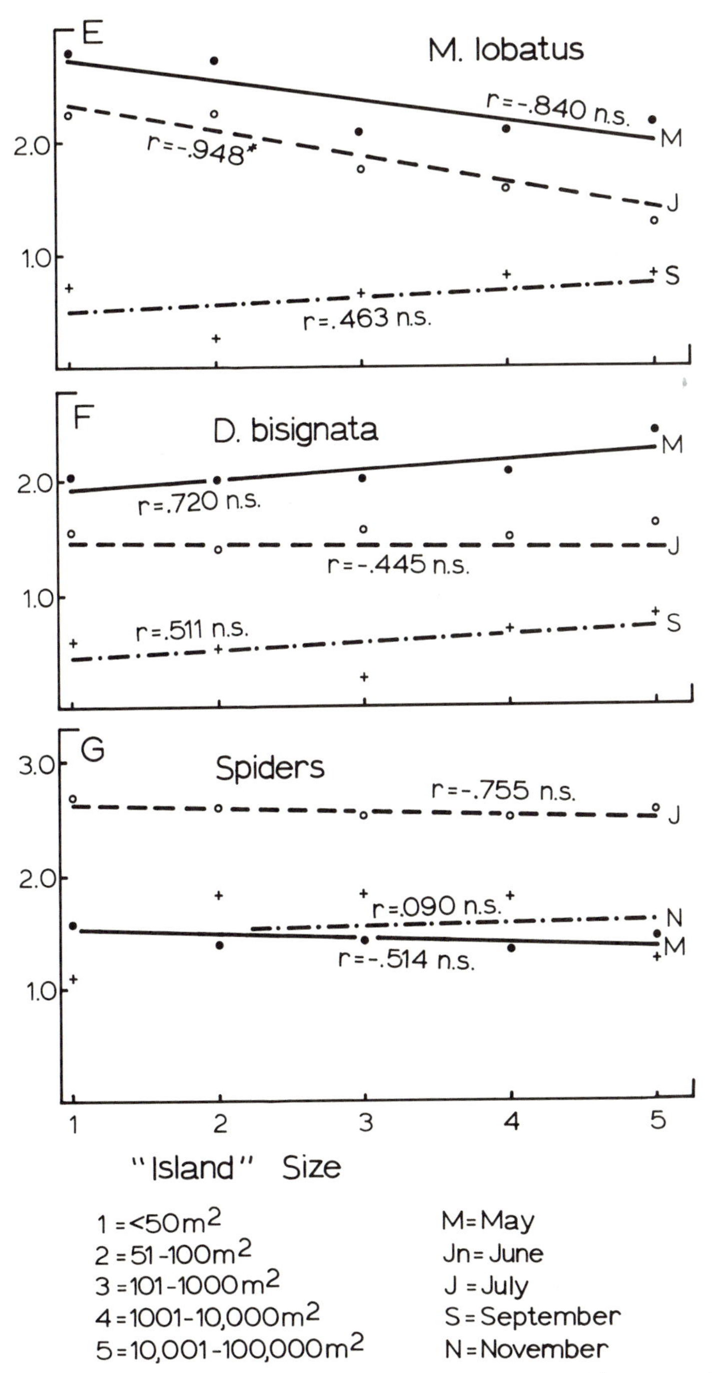

Figure 9-11. Relationship between the "island" (stand) size of *S. patens* and the densities of sap-feeders and spiders at several times during 1979 at Tuckerton, New Jersey. Means are based on 4 samples taken in 3 "islands" in each size category (largest size excepted) with a D-vac vacuum sampler. One sample consisted of 4 placements (30 sec each) of the sampler on the *S. patens* surface. Only one "island" in the 10,001-100,000 m^2 category occurred at the study area (** $P < 0.01$; * $P < 0.05$; n.s., not significant).

Species with the opposite combination of life history traits (e.g., high colonization ability coupled with reduced winter extinction rate) should have the best chance for success on small islands. The only species with this combination of traits, *Am. simplex*, was equally abundant on all sizes of *S. patens* patches during May and June and even showed a significant negative relationship between density and "island" size during July (Fig. 9-11C).

Species with good colonizing ability (10-20% macropters) and exposed overwintering stages like *D. detecta,* or species with poor colonizing ability ($>$ 95% brachypters) and protected overwintering stages like *M. lobatus* and *D. bisignata,* were for the most part equally abundant on small and large "islands," but for different reasons (Fig. 9-11D, E and F). *D. detecta,* even though it was absent from small "islands" (< 50 m^2) during May, achieved densities later in the season (November) similar to those in the larger island sizes (Fig. 9-11D). Winter mortality is great on small islands, but macropterous adults colonize these sites early in the season and subsequently reproduce (Denno et al., in prep.).

By contrast, *M. lobatus* (97% brachypterous) and *D. bisignata* (100% of females brachypterous) have poor colonizing ability, but withstand the rigors of winter very well in both small and large patches of grass because they embed their eggs. Compare May densities (populations composed of newly hatched nymphs at this time) of these species on large and small islands (Fig. 9-11E and F). Once they have colonized small islands, extinction rates for these species are apparently low. Colonization may be a rare event, but we emphasize that nymphs and adults of these species are able to drift on the tidewater's surface by virtue of their waxy integument. However, the mortality associated with drifting behavior must be great, particularly during storm and high spring tides, due to the inability of these insects to direct their course on the water's surface and greater susceptibility to density-independent mortality factors. It seems unlikely that selection would favor such precarious dispersal behavior over macroptery in adults if dispersal were an adaptive event in the life histories of *M. lobatus* and *D. bisignata.* Instead, we suggest that the life histories of these two species have been shaped to ensure intimate contact with immediate large patches of grass and that their occupation of small islands is the result of historical accident.

To ensure that the density responses of the various sap-feeders to grass-patch size did not constitute a reflection of predation, we also measured spider density along the *S. patens* "island" size gradient. Spiders, mostly in the family Lycosidae, are by far the most abundant predators in the *S. patens* marsh system (Denno 1977, Raupp and Denno 1979). Measured during May, July and November, there was no significant relationship between spider density and "island" size (Fig. 9-11G). Parasitoids of the eggs, nymphs and adults of these sap-feeders are not abundant in the mid-Atlantic states (Denno 1977). Therefore, it is unlikely that the responses of several of the sap-feeders to patch size are attributable to either predation or parasitism.

Lastly, we emphasize that the life histories of all but one sap-feeder, the fully winged leafhopper *Am. simplex,* have been adjusted in varying degrees to maintain contact with occupied patches of *S. patens.* The brachypterous adults and/or embedded eggs characteristic of most species are consistent with this hypothesis. At high densities, some planthoppers are able to alter their life history by virtue of a developmental switch and long-winged morphs are produced. This behavior can be adaptive only if more favorable alternative sites are available. The argument can be made that small "islands" of *S. patens* are more favorable than large ones because certain species (potential competitors) are rare there (see Fig. 9-11A and B). However, small patches have less protective thatch than large patches and are probably less favorable, particularly for those species that overwinter as an exposed stage. Thus, small "islands" probably serve as predictable refuges for only *Am. simplex*. The presence of other sap-feeders on small patches is due probably to either accidental colonization by nymphs or dispersal among larger patches by adults. Consequently, nonequilibrium conditions are probably the rule rather than the exception on small patches. In the section that follows we show that large patches too can be even more favorable alternatives for colonization because they are sporadically free of potential competitors, yet unlike small "islands" have more protective thatch.

7 Equilibrium Status of the Sap-feeder Guild in Large Patches

A six-year assessment of sap-feeders on the same large patches (> 10 ha) of *S. patens* in New Jersey reveals drastic fluctuations in the populations of some species and not others. For instance, early-season (May, June) densities of *D. detecta* during 1978 and 1979 were excessively low compared to the previous four years, while *Am. simplex* failed to show this decline (Fig. 9-12). A more detailed analysis shows that sap-feeder species fit into one of two categories. There are three species, *D. detecta, T. minuta* and *A. simplex,* whose populations were drastically reduced following the winters of 1978-79, and particularly 1977-78 (Fig. 9-13A, B and C). Three other species, *Am. simplex, M. lobatus* and *D. bisignata,* did not show this pattern, and in fact their early-season densities were highest following the winter of 1977-78, when the densities of the other three species were most depressed (Fig. 9-13D, E and F). We anticipated

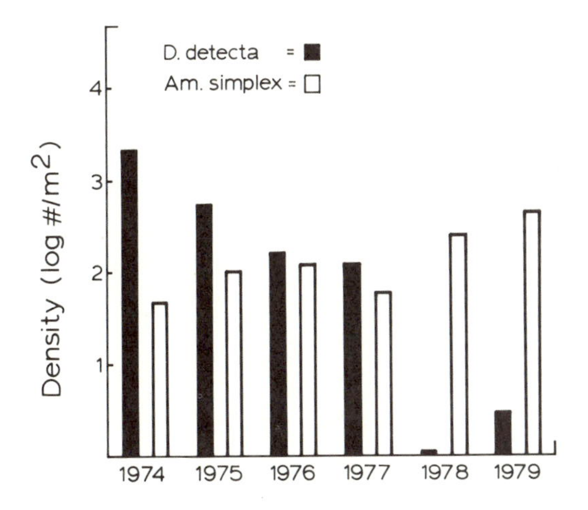

Figure 9-12. Mean spring densities (May-June) of *D. detecta* and *Am. simplex* in large stands (> 1000 m^2) of *S. patens* at Tuckerton, New Jersey, 1974-1979.

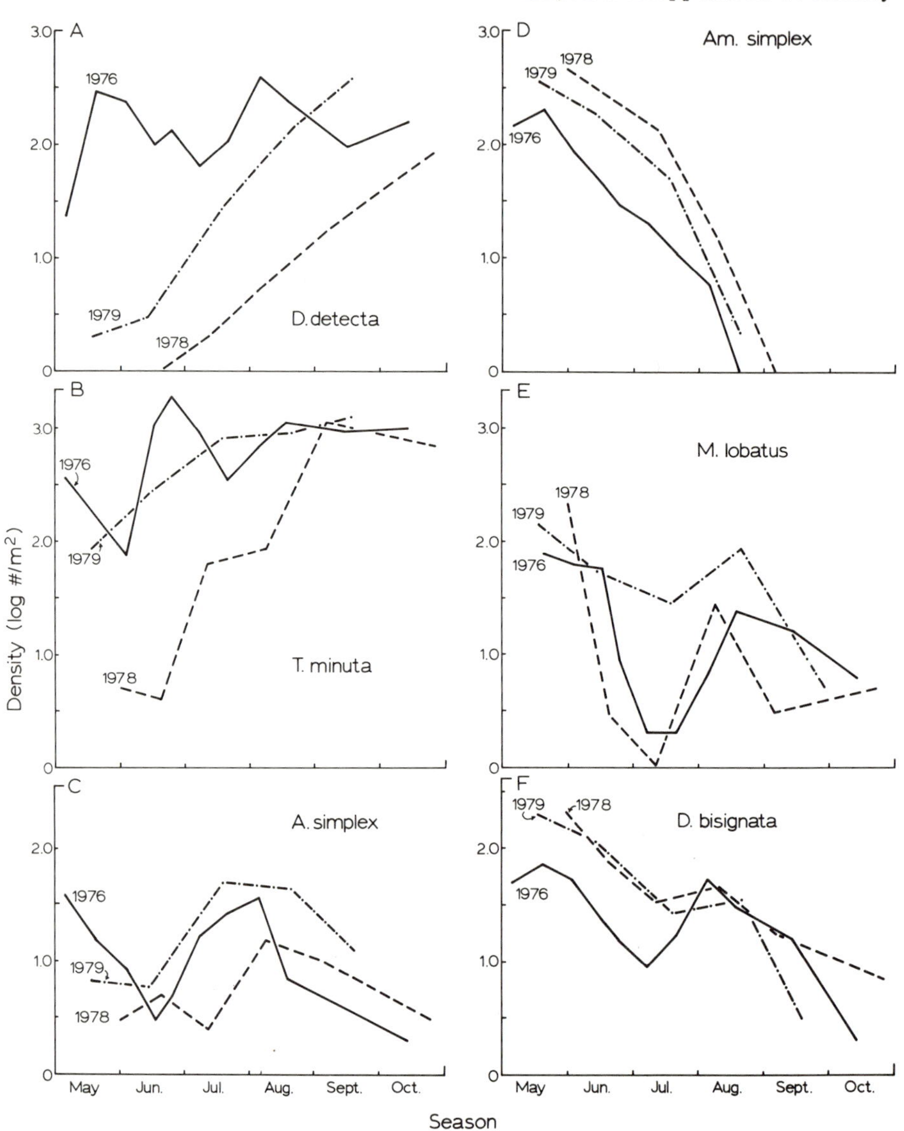

Figure 9-13. Seasonal densities of sap-feeders in a large stand (> 10,000 m^2) of *S. patens* during 3 years at Tuckerton, New Jersey. Winters of 1977-78 and 1978-79 were severe compared to that of 1975-76. Insects were sampled with a D-vac vacuum as in Fig. 9-11.

that severe winters were selectively killing *D. detecta, T. minuta* and *A. simplex,* but not the remaining species.

The relationship between the severity of the previous winter (number of days the minimum daily temperature fell below 0 °C) and the following early-season density (measured prior to reproduction) was negative and highly significant for these three species substantiating the winter-kill hypothesis (Fig. 9-14A, B and C). Regressions

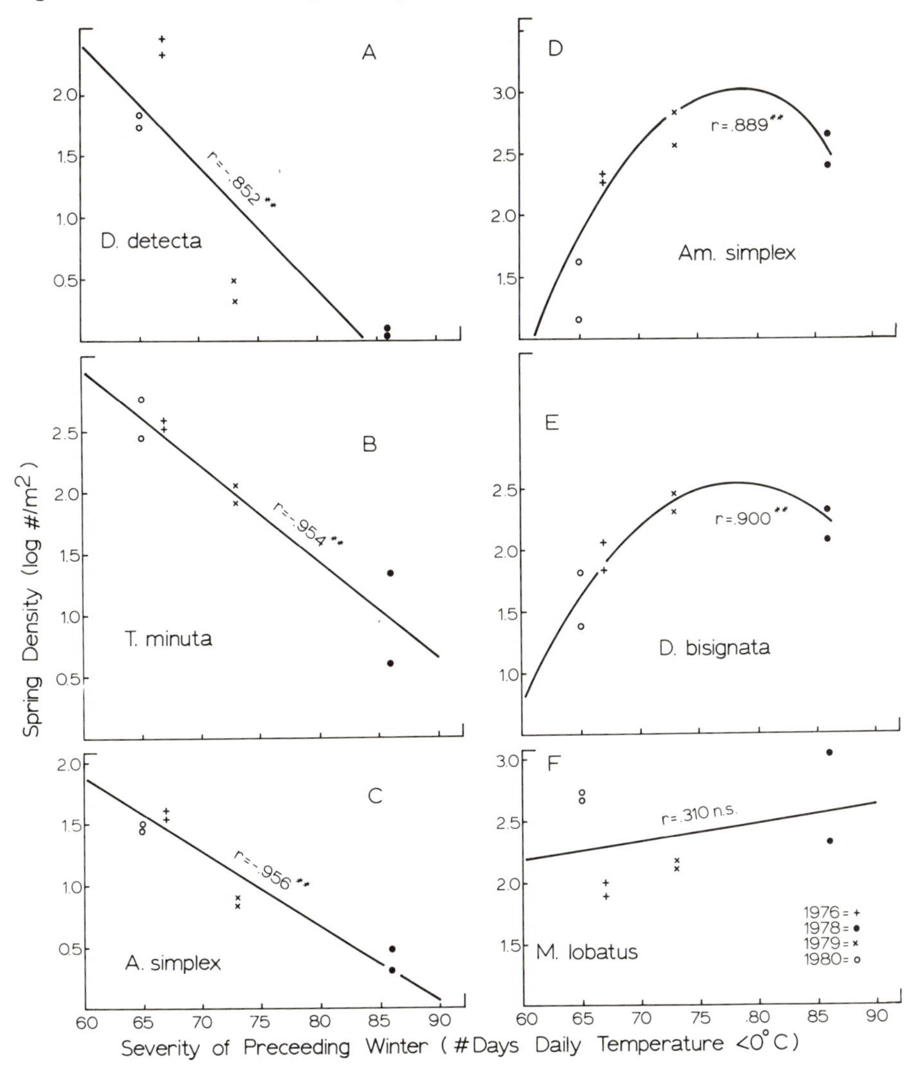

Figure 9-14. Relationship between the severity of the preceding winter (no. of days the daily temperature fell below 0 °C from April 1 to October 31) and the following spring density of sap-feeders (May or June density prior to reproduction). Insect samples were taken with a D-vac vacuum as in Fig. 9-11 during a 4-year period (1976-1980) in 2 large stands (> 10,000 m^2) of *S. patens* at Tuckerton, New Jersey.

using average daily temperature during the winter season and number of days the maximum daily temperature fell below 0 °C as other measures of winter harshness (independent variables) produced the same results. There was no relationship between the severity of the previous winter and the spring density of *M. lobatus* (Fig. 9-14F). *Am. simplex* and *D. bisignata* showed a significant positive relationship between winter severity and spring density (Fig. 9-14D and E). The only explanation we have to offer for this result is that predators (spiders and the mirid egg predator, *Tytthus alboornatus*

Knight) may be more active during mild winters, resulting in higher mortality on those years.

It is not surprising that the species which were affected by winter-kill all overwinter as either exposed nymphs or eggs. Species not affected by severe winters all overwinter as embedded eggs in the vegetation. The result is a guild of sap-feeders, half of which can be brought to near local extinction following very harsh winters.

These data conjure up questions concerning the equilbrium status of the guild, the degree and frequency of resource limitation, and the reality of competition among the guild members. Following a severe winter (1978), certain species (see *T. minuta,* Fig. 9-13B) reach densities by the end of the season that compare with those after mild winters (e.g., 1976). Other species (*D. detecta* and *A. simplex,* Fig. 9-13A and C) fail to attain densities that they obtain during the season following a mild winter. Thus, certain species appear to be contained well below carrying capacity at a frequency proportional to the occurrence of severe winters. Consequently large patches of grass as well as small ones can be relatively empty in certain years, apparently allowing for the coexistence of competitively weaker species under nonequilibrium conditions.

8 The Fugitive Coexistence of *Amplicephalus simplex*

Based on our niche analysis, *Am. simplex* appears to occupy a rather precarious position in the guild of sap-feeders (see Fig. 9-7). It shares a niche very similar to that of *A. simplex* as well as *D. detecta,* and it appears to be negatively affected by the presence of these species (Fig. 9-8A and C). We do not think it is coincidental that, of the six sap-feeder species on the low marsh in the mid-Atlantic area, *Am. simplex* is the only one totally retaining its ability to fly. Flightlessness is a common life history feature in the remaining five species.

Vacant sites are frequently available on small "islands" of *S. patens,* offering a rather consistent refuge for *Am. simplex,* and *A. simplex* is always rare there (Fig. 9-11B). Also, relatively vacant sites are unpredictably created in large patches by the catastrophic reduction of its closest competitors during severe winters (see *D. detecta* and *A. simplex,* Fig. 9-14A and C). Our results are consistent with the hypothesis that *Am. simplex* retains its wings and hedges its bets by moving among patches of *S. patens* depositing some eggs in each. In New Jersey frequent disturbances apparently allow this species to coexist in the guild.

9 Latitudinal Distribution of Sap-feeders

One prediction emerging from the previous discussion is that, as the frequency of disturbance (winter-kill of competitors) decreases, so should the probability of *Am. simplex* coexisting in the guild. A latitudinal transect along the eastern seaboard of North America from Hampton Beach, New Hampshire, to Crescent Beach, Florida, confirms this prediction (Fig. 9-15). South of Virginia, *Am. simplex* does not occur on *S. patens* and becomes increasingly more abundant to the north (Fig. 9-16E). Also consistent

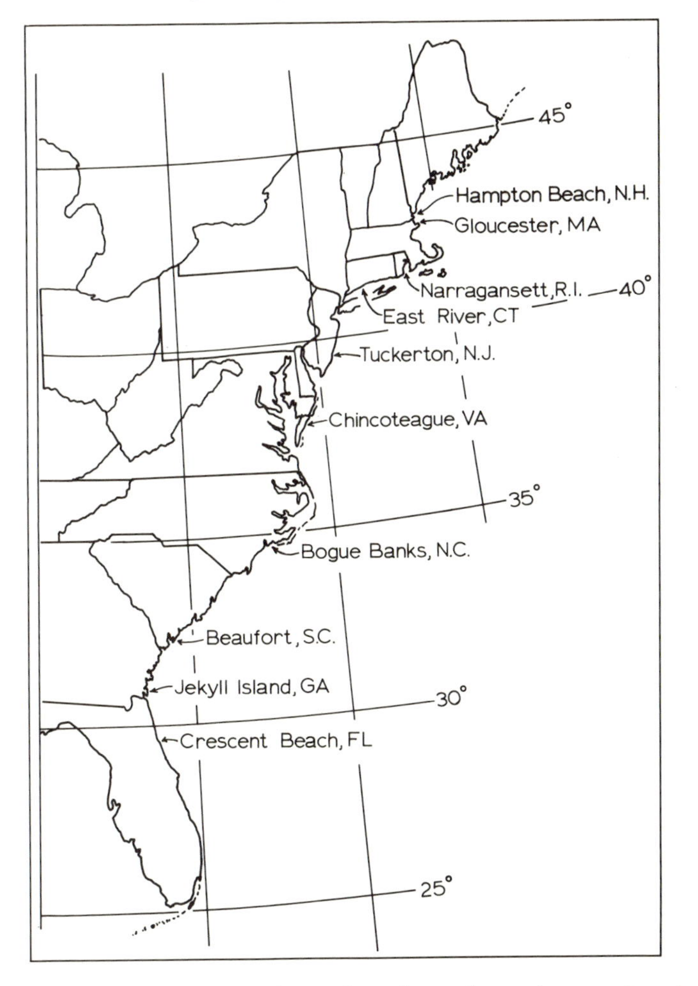

Figure 9-15. Locations of intertidal marshes along the eastern seaboard of North America sampled to determine latitudinal distributions of sap-feeders on *S. patens*.

with the argument is the fact that *D. detecta* and the *A. simplex-A. decorata* complex (species that share a niche very similar to *Am. simplex*) become increasingly more abundant on southern marshes (Fig. 9-16A and B).

Latitudinal species patterns appear to be mediated in part by life history type that is legislated by climate. For instance, univoltine species that embed their eggs in vegetation (*Am. simplex* and *H. lineatus*) occur only on northern marshes (Fig. 9-16E and F). It may be that univoltine sap-feeders are disadvantaged at southern latitudes because of long periods of exposure to competitors, egg predators and/or parasitoids coupled with infrequent bouts of reproduction. The positive relationship we found between the severity of the preceding winter and the following early-season density of *Am. simplex* is consistent with the argument that mortality is greater when there is potentially greater exposure to predators (Fig. 9-14D).

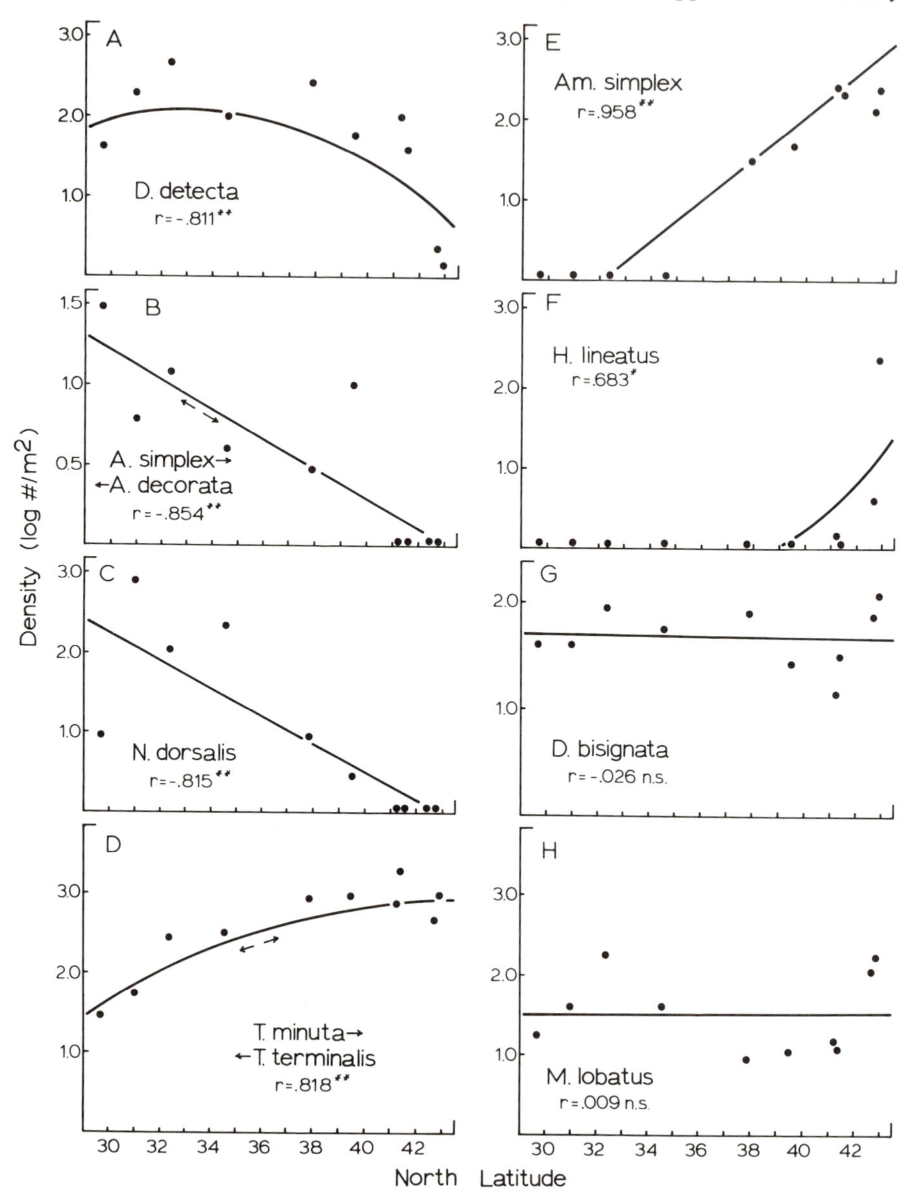

Figure 9-16. Relationship between density and latitude for sap-feeders on *S. patens*. Sample points correspond to the 10 locations from northern Florida to New Hampshire shown in Fig. 9-15. Points are means calculated from 6 samples taken in stands of *S. patens* at least 500 m^2 in area. One sample as in Fig. 9-11. Marshes to Chincoteague, Virginia south, were sampled during early June 1980. Tuckerton, New Jersey, and locations north were sampled during early July 1980 (** $P < 0.01$; * $P < 0.05$; n.s., not significant).

Southern marshes are dominated by multivoltine species that overwinter as either exposed eggs (*A. simplex*) or nymphs (*D. detecta* and *N. dorsalis*) (Fig. 9-16A, B and C). These species are apparently susceptible to cold temperatures and are selectively eliminated at northern latitudes (see Fig. 9-14A, B and C). However, farther to the south where winter temperatures are less severe, it may adaptive to overwinter as nymphs that are able to begin feeding as soon as temperature permits in spring, when the quality of the grass is highest (see Fig. 9-2). Mittler (1958), van Emden (1966), Dixon (1969), Auclair (1976), Horsfield (1977), McClure (1977, 1979) and Mitsuhashi and Koyama (1977) have all shown that the fitness (fecundity, growth rate, survival and adult size) of sap-feeding insects is positively related to host plant quality (total nitrogen or amino nitrogen). The *T. minuta-T. terminalis* complex is also multivoltine and overwinters as nymphs in New Jersey, but is more abundant on northern marshes (Fig. 9-16D). There may be several reasons for this apparent exception. First, of the multivoltine species that overwinter as exposed stages, *T. minuta* is the only species that inhabits the lower stratum of the grass system (see Fig. 9-4); *D. detecta, N. dorsalis* and *A. simplex* all reside in the upper stratum (see Fig. 9-3). This may afford more protection, particularly during harsh winters. The fact that *T. minuta* was present on the marsh in June 1978 while *D. detecta* was not suggests that *T. minuta* is better able to survive severe winter temperatures (compare Fig. 9-13A with B). An alternative explanation for *T. minuta*'s success in the north may be a variable life history. While it overwinters as susceptible nymphs in New Jersey, it has the potential to overwinter as embedded eggs further to the north. *T. minuta* does have a saw-like ovipositor very much like that of *M. lobatus* (see Fig. 9-10). We would add, however, that *D. detecta* also have a saw-like ovipositor and thus the potential to vary its life history, but is rare or absent from the northernmost marshes (Fig. 9-16A). Thus, it seems likely that the association of *T. minuta* with the protective thatch of *S. patens* offers the most parsimonious explanation for its success on northern marshes.

Bivoltine species that overwinter as embedded eggs in New Jersey (*D. bisignata* and *M. lobatus*) were similarly abundant along the entire latitudinal gradient (Fig. 9-16G and H). It may be that these species vary their life history with latitude, overwintering as eggs to the north and reproducing continuously in the south. The presence of adults of these species during April, June, July and late October support this contention. If this is the case, the advantage of embedded eggs to the north and continuous reproduction to the south might explain the success of *D. bisignata* and *M. lobatus* on *S. patens* from Florida to New Hampshire. *D. bisignata* also shows depressed densities following mild winters in New Jersey, suggesting that it may incur increased mortality from egg predators there (Fig. 9-14E). This latter possibility underscores the importance of continuous reproduction to the south, unless there are other overriding constraints like those associated with host plant quality or quantity.

10 Summary and Conclusions

Population equilibrium exists when birth and death rates are equal, resulting in a net growth rate of zero (MacArthur and Wilson 1967, May 1973, Price 1980). More practically, equilibrium exists when population size fluctuates with a steady average variance around an average population size (May 1973, Price 1980). Price (1980) defines a population in a nonequilibrium state as one that does not fluctuate within a typical probability range around an average population size. He also suggests that steady population growth following colonization, with subsequent rapid extinction, provides no indication of an equilbrium. The populations of several of the sap-feeders on *S. patens* fluctuate in ways that strongly violate the tenets of equilibrium theory and suggest nonequilibrium behavior. For instance, following the severe winter of 1977-78, populations of *A. simplex* and *T. minuta* were drastically reduced, and *D. detecta* was annihilated altogether from the study area in Tuckerton, New Jersey (Fig. 9-13A, B and C). Furthermore, we show that in general the overwintering success of these species is negatively related to the severity of the winter (Fig. 9-14A, B and C). Other species are unaffected by severe winters (Figs. 9-13 and 9-14D, E and F). Consequently, certain species can be selectively reduced or eliminated from the guild following a harsh winter and never attain the population size at the end of the season that they do following mild winters (e.g., Fig. 9-13A).

The overwintering success of a species is dictated in large part by the exposure of the overwintering stage. Species like *D. detecta* and *A. simplex* that overwinter as exposed nymphs and eggs respectively incur great mortality during severe winters compared to species like *Am. simplex,* which protect their eggs by inserting them into vegetation. Where winters are usually severe in northern New England, most species that overwinter in an exposed stage are either absent or extremely rare (Fig. 9-16A, B and C).

Southern marshes, where mild winters prevail, are characterized by multivoltine species that overwinter as exposed nymphs (*D. detecta*), exposed eggs (*A. simplex*) or embedded eggs (*D. bisignata*). Univoltine species like *Am. simplex* and *H. lineatus,* which are so common on northern marshes, are absent on southern marshes. We suggest that a high premium is placed on continuous reproduction and development in the south, where exposure of a single stage to competitors, predators and parasitoids is minimized.

In mid-Atlantic states (New Jersey), and probably in southern New England states as well, where the variance in winter harshness is great, we suggest that populations of *D. detecta* and *A. simplex* in particular are maintained below carrying capacity at a frequency equivalent to the occurrence of catastrophic winters. Consequently, populations of these species in both large and small patches are often in a state of nonequilibrium. Additionally, certain species of sap-feeders are consistently rare or absent on small patches, suggesting that they are inherently harsher and harder to track than large patches. Thus, the picture we paint is one of frequent nonequilibrium conditions in both space and time.

Despite our arguments in support of a nonequilibrium hypothesis, the guild of sap-feeders associated with *S. patens* appears to be organized in a way that is also consistent with a competition interpretation. There is a considerable amount of niche diversification, particularly along the microhabitat dimension, where species sort out into upper- and lower-strata residents. Also, among the upper-stratum species there are considerable negative density effects. At our major study site in New Jersey, a niche analysis suggests that the sap-feeder guild is overpacked where *Am. simplex, A. simplex* and *D. detecta* share very similar niches. Of these species and of the remainder as well, *Am. simplex* is the only one that has fully developed wings and can fly. We suggest that this species can coexist in the guild by colonizing empty sites created by cold-winter elimination of its closest competitors, *A. simplex* and *D. detecta*. South of New Jersey, where cold-winter disturbance is rare and competitors persistent, *Am. simplex* is rare and drops out completely south of Virginia. To the north of New Jersey *Am. simplex* is abundant.

The structure of the subguild of species associated with the lower stratum of the grass remains unchanged from Florida to New Hampshire (*D. bisignata, M. lobatus* and the *T. minuta-T. terminalis* complex are similarly abundant along the entire latitudinal gradient). The buffering effects of the thatch coupled with overwintering style may be responsible for the constancy in structure of this subguild. There are several latitudinal replacements among upper-stratum species. *Am. simplex* and *H. lineatus* occur on New England marshes, but are absent on southern marshes, where *N. dorsalis* and the *A. simplex-A. decorata* complex are restricted. All of these species occur on mid-Atlantic state marshes, but *H. lineatus* and *N. dorsalis* are rare there (Table 9-4). We suggest that levels of disturbance (winter mortality) are intermediate on mid-Atlantic compared to southern and New England marshes, and that sap-feeder richness is highest there because the populations of potential competitors are often in a state of nonequilibrium.

We have tried to interpret our data according to both nonequilibrium and equilibrium hypotheses. To date, two features of our system impress us. First, density-independent mortality coupled with patch heterogeneity appear to play a major role in preventing the populations of certain species from often attaining equilibrium numbers and in shaping the life histories of those species. Second, if our data are correct in suggesting that equilibrium conditions are attained often enough to maintain niche diversification among the various sap-feeders, then competition must be intense during at least a portion of the time these conditions occur. Finally, the validity of a competition hypothesis will be elucidated only by laboratory experiments where the fitness of sap-feeders is evaluated in paired compared to pure cultures established at realistic field densities. Only then will we be able to assess the impact of species interactions in guilds of herbivorous insects.

Table 9-4. Richness of sap-feeders on New England, mid-Atlantic and southern marshes

Southern	Mid-Atlantic	New England	Stratum
T. minuta-T. terminalis	*T. minuta-T. terminalis*	*T. minuta-T. terminalis*	lower
D. bisignata	*D. bisignata*	*D. bisignata*	lower
M. lobatus	*M. lobatus*	*M. lobatus*	lower
D. detecta	*D. detecta*	*D. detecta* (rare)	upper
A. simplex-A. decorata	*A. simplex-A. decorata*		upper
N. dorsalis	*N. dorsalis* (rare)		upper
	Am. simplex	*Am. simplex*	upper
	H. lineatus (rare)	*H. lineatus*	upper
Species total 6	8	6	

Acknowledgments. Barbara Denno processed the many samples for this study. Vera Krischik, Mark McClure, Peter Price, Don Strong and Tom Whitham made helpful comments on earlier drafts of this report. Jim Kramer (Systematic Entomology Laboratory, USDA) made the initial sap-feeder identifications. Susan Smith typed the several drafts of this report. To these people we are most grateful.

Scientific Article No. A2888, Contribution No. 5942 of the Maryland Agricultural Experiment Station, Department of Entomology. Figs. 9-3 to 9-8 were reproduced, with permission, from Denno (1980), Duke University Press, Durham, N.C., U.S.A.

References

Adams, D. A.: Factors influencing vascular plant zonation in North Carolina salt marshes. Ecology 44, 445-456 (1963).

Auclair, J. L.: Feeding and nutrition of the pea aphid, *Acyrthosiphon pisum* (Harris), with special reference to amino acids. Symp. Biol. Hung. 16, 29-34 (1976).

Blum, J. L.: Salt marsh spatinas and associated algae. Ecol. Monogr. 38, 199-221 (1968).

Brinkhurst, R. O.: Observations on wing-polymorphism in the Heteroptera. Proc. Roy. Entomol. Soc. Lond. 38, 15-22 (1963).

Brown, J. H.: Geographical ecology of desert rodents. In: Ecology and Evolution of Communities. Cody, M. L., Diamond, J. M. (eds.). Cambridge, Mass.: Belknap Press, 1975, pp. 315-341.

Caswell, H.: Predator-mediated coexistence: A non-equilibrium model. Am. Nat. 112, 127-154 (1978).

Cody, M. L.: Competition and the structure of bird communities. Princeton, N.J.: Princeton Univ. Press, 1974.

Connell, J. H.: The influence of interspecific competition and other factors on the distribution of the barnacle *Chthamalus stellatus.* Ecology 42, 710-723 (1961).

Connell, J. H.: Diversity in tropical rain forests and coral reefs. Science 199, 1302-1309 (1978).

Davis, L. V., Gray, I. E.: Zonal and seasonal distribution of insects in North Carolina salt marshes. Ecol. Monogr. 36, 275-295 (1966).

Dayton, P. K.: Competition, disturbance, and community organization: The provision and subsequent utilization of space in a rocky intertidal community. Ecol. Monogr. 41, 351-389 (1971).

Denno, R. F.: Ecological significance of wing polymorphism in Fulgoroidea which inhabit tidal salt marshes. Ecol. Entomol. 1, 257-266 (1976).

Denno, R. F.: Comparison of the assemblages of sap-feeding insects (Homoptera-Hemiptera) inhabiting two structurally different salt marsh grasses in the genus *Spartina.* Environ. Entomol. 6, 359-372 (1977).

Denno, R. F.: The optimum population strategy for planthoppers (Homoptera: Delphacidae) in stable marsh habitats. Can. Entomol. 110, 135-142 (1978).

Denno, R. F.: The relation between habitat stability and the migration tactics of planthoppers. Misc. Publ. Entomol. 11, 41-49 (1979).

Denno, R. F.: Ecotope differentiation in a guild of sap-feeding insects on the salt marsh grass, *Spartina patens.* Ecology 61, 702-714 (1980).

Denno, R. F., Grissell, E. E.: The adaptiveness of wing-dimorphism in the salt marsh-inhabiting planthopper, *Prokelisia marginata* (Homopetera: Delphacidae). Ecology 60, 221-236 (1979).

Denno, R. F., Raupp, M. J., Tallamy, D. W., Reichelderfer, C. F.: Migration in heterogeneous environments: Differences in habitat selection between the wing-forms of the dimorphic planthopper, *Prokelisia marginata* (Homptera: Delphacidae). Ecology 61, 859-867 (1980).

Diamond, J. M.: Niche shift and the rediscovery of interspecific competition. Am. Sci. 66, 322-331 (1978).

Dixon, A. F. G.: Quality and availability of food for a sycamore aphid populations. In: Animal Populations in Relation to their Food Resources. Watson, A. (ed.). Oxford: Blackwell Scientific Publ., 1969, pp. 271-287.

Gause, G. F.: The Struggle for Existence. Baltimore, Md.: Williams and Wilkins, 1934.

Hair, J. D., Holmes, J. C.: The usefulness of measures of diversity, niche width and niche overlap in the analysis of helminth communities of waterfowl. Acta Parasitol. Pol. 23, 253-269 (1975).

Harper, J. L.: Population Biology of Plants. New York: Academic Press, 1977.

Holmes, J. C.: Site selection by parasitic helminths: Interspecific interactions, site segregation, and their importance to the development of helminth communities. Can. J. Zool. 51, 333-347 (1973).

Horsfield, D.: Relationships between feeding of *Philaenus spumarius* (L.) and the amino acid concentration in the xylem sap. Ecol. Entomol. 2, 259-266 (1977).

Horwitz, W.: Methods of Analysis of the Association of Official Agricultural Chemists. Washington, D.C.: Assoc. Agric. Chemists, 1965.

Hubbell, S. P.: Tree dispersion, abundance, and diversity in a tropical dry forest. Science 203, 1299-1309 (1979).

Hutchinson, G. E.: The concept of pattern in ecology. Proc. Acad. Nat. Sci. USA 105, 1-12 (1953).

Hutchinson, G. E.: Homage to Santa Rosalia or why are there so many kinds of animals? Am. Nat. 93, 145-159 (1959).

Hutchinson, G. E.: An Introduction to Population Ecology. New Haven, Conn.: Yale Univ. Press, 1977.

Johnson, C. G.: Migration and Dispersal of Insects by Flight. London: Methuen, 1969.

Kennedy, C. R.: The regulation of fish parasite populations. In: Regulation of Parasite Populations. Esch, G. W. (ed.). New York: Academic Press, 1977, pp. 63-109.

Kisimoto, R.: Factors determining the wing-form of adult, with special reference to the effect of crowding during the larval period of the brown planthopper, *Nilaparvata lugens* Stål. Studies on the polymorphism in the planthoppers (Homoptera, Areopidae). I. Ōyō-Kontyū 12, 105-111 (1956).

Kisimoto, R.: Studies on the polymorphism and its role playing in the population growth of the brown planthopper, *Nilaparvata lugens* Stål. Bull. Shikohu Agric. Exp. Stn. 13, 1-106 (1965).

Lotka, A. J.: The structure of a growing population. Hum. Biol. 3, 459-493 (1931).

MacArthur, R. H.: Geographical Ecology–Patterns in the Distribution of Species. London: Harper and Row, 1972.

MacArthur, R. H., Wilson, E. O.: The Theory of Island Biogeography. Princeton, N.J.: Princeton Univ. Press, 1967.

May, R. M.: Stability and Complexity in Model Ecosystems. Princeton, N.J.: Princeton Univ. Press, 1973.

May, Y. Y.: The biology and population ecology of *Stenocranus minutus* (Fab.) (Delphacidae, Hemiptera). Ph.D. Dissertation, University of London, 1971.

McClure, M. S.: Population dynamics of the red pine scale, *Matsucoccus resinosae* (Homoptera: Margarodidae): The influence of resinosis. Environ. Entomol. 6, 789-795 (1977).

McClure, M. S.: Self-regulation in hemlock scale populations: Role of food quantity and quality. Misc. Publ. Entomol. 11, 33-39 (1979).

McClure, M. S., Price, P. W.: Competition and coexistence among sympatric *Erythroneura* leafhoppers (Homoptera: Cicadellidae) on American sycamore. Ecology 56, 1388-1397 (1975).

McClure, M. S., Price, P. W.: Ecotype characteristics of coexisting *Erythroneura* leafhoppers (Homoptera: Cicadellidae) on sycamore. Ecology 57, 928-940 (1976).

Miller, W. R., Egler, F. E.: Vegetation of the Wequetequock-pawcatuck tidemarshes, Connecticut. Ecol. Monogr. 20, 143-172 (1950).

Mitsuhashi, J., Koyama, K.: Effects of amino acids on the oviposition of the Smaller Brown Planthopper, *Laodelphax striatellus* (Hemiptera: Delphacidae). Entomol. Exp. Appl. 22, 156-160 (1977).

Mittler, T.: Studies on the feeding and nutrition of *Tuberolachnus salignus* (Gmelin) (Homoptera: Aphididae). II. The nitrogen and sugar composition of ingested phloem sap and excreted honeydew. J. Exp. Biol. 35, 74-84 (1958).

Mobberley, D. G.: Taxonomy and distribution of the genus *Spartina*. J. Sci. Iowa St. Coll. 30, 471-574 (1956).

Mochida, O.: The characters of the two wing-forms of *Javesella pellucida* (F) (Homoptera: Delphacidae), with special reference to reproduction. Trans. Roy. Entomol. Soc. Lond. 125, 177-225 (1973).

Nasu, S.: The Virus Diseases of the Rice Plant. Baltimore: Johns Hopkins Press, 1969.

Price, P. W.: Evolutionary Biology of Parasites. Princeton, N.J.: Princeton Univ. Press, 1980.

Raupp, M. J., Denno, R. F.: The influence of patch size on a guild of sap-feeding insects that inhabit the salt marsh grass *Spartina patens*. Environ. Entomol. 8, 412-417 (1979).

Redfield, A. C.: Development of a New England salt marsh. Ecol. Monogr. 42, 201-237 (1972).

Rey, J. R.: Ecological biogeography of arthropods on small islands in northwest Florida. Ecology (1981) (in press).

Rosenzweig, M. L., Sterner, P.: Population ecology of desert rodent communities: Body size and seed husking as bases for heteromyid coexistence. Ecology 51, 217-224 (1970).

Schoener, T. W.: Resource partitioning in ecological communities. Science 185, 27-39 (1974).

Simberloff, D. S.: Experimental zoogeography of islands: A model for insular colonization. Ecology 50, 296-314 (1969).

Simberloff, D. S.: Experimental zoogeography of islands: Effects of island size. Ecology 57, 629-648 (1976).

Simberloff, D., Boecklen, W.: A bum rap for Santa Rosalia: Size ratios and competition. Am. Nat. (1981) (in press).

Simberloff, D. S., Wilson, E. O.: Experimental zoogeography of islands: The colonization of empty islands. Ecology 50, 278-296 (1969).

Simberloff, D. S., Wilson, E. O.: Experimental zoogeography of islands: A two-year record of colonization. Ecology 51, 934-937 (1970).

Skellam, J. G.: Random dispersal in theoretical populations. Biometrica 38, 196-218 (1951).

Slobodkin, L. B.: Growth and Regulation of Animal Populations. New York: Holt, Reinhart and Winston, 1961.

Southwood, T. R. E.: Migration of terrestrial arthoropods in relation to habitat. Biol. Rev. 37, 171-214 (1962).

Strong, D. R.: Insect species richness: hispine beetles of *Heliconia latispatha*. Ecology 58, 573-582 (1977a).

Strong, D. R.: Rolled-leaf hispine beetles (Chrysomelidae) and their Zingiberales host plants in Middle America. Biotropica 9, 156-169 (1977b).

Tallamy, D. W., Denno, R. F.: Responses of sap-feeding insects (Homoptera-Hemiptera) to simplification of host plant structure. Environ. Entomol. 8, 1021-1028 (1979).

Tsai, P., Hwang, F., Feng, W., Fu, Y., Dong, Q.: Study on *Delphacodes striatella* Fallen (Homoptera, Delphacidae) in north China. Acta Entomol. Sin. 13, 552-571 (1964).

van Emden, H. F.: Studies on the relations of insect and host plant. III. A comparison of the reproduction of *Brevicoryne brassicae* and *Myzus persicae* (Hemiptera: Aphididae) on Brussels sprout plants supplied with different rates of nitrogen and potassium. Entomol. Exp. Appl. 9, 444-460 (1966).

Chapter 10

The Possibility of Insect Communities without Competition: Hispine Beetles on *Heliconia*

DONALD R. STRONG

1 Introduction

Much ecological theory is based upon neo-Malthusian population ecology, which assumes that competitive interactions among species are the preeminent cause of structure in communities (Hutchinson 1977). Usually, this competition has been observed by ecologists working with vertebrates (MacArthur 1972), plants (Harper 1977), or intertidal marine organisms (Paine 1974). Entomologists, on the other hand, have less frequently included interspecies among the more important causes of community structure. This raises the fascinating question of whether insect communities are formed and maintained by forces distinct from those operating among vertebrates, plants and benthic organisms, or whether a close look will reveal similar community mechanisms in all groups. Insects are the most speciose organisms, and any consistent distinctions in their community structure are of great interest to ecology.

One view holds that "the tropics (are) where the evidence for competition is clearest" (Diamond 1978). With this in mind I have studied a group of phytophagous insects, the rolled-leaf hispine beetles. Adults of up to six species of these chrysomelids live together in the scrolls of immature, rolled leaves of the monocot *Heliconia,* where they feed and mate (Strong 1977a, 1977b). These beetles are always either in a rolled leaf or searching for one. The only exceptions are two species that live and feed in inflorescences during the flowering season of some *Heliconia* species. Scrolls are loose enough to accommodate beetles for from a few days to several weeks before they unfurl, depending upon local conditions and the *Heliconia* species. Leaves are large, and, when they loosen, space and leaf tissue for food are abundant in the scroll. Hispines, together with all other phytophages, only rarely eat as much as 0.5% of the leaf while it is rolled. Experiments show that much remaining tissue is as desirable as food as that which is consumed (Strong, unpubl.). *Heliconia* stalks bear only a single rolled leaf at a time and have a long period between rolled leaves, so beetles must fly among rolled leaves. A leaf has both arrivals and departures throughout the time that it accommodates beetles, so that the insects are constantly resorted among the rolled leaves of a host population. All life history stages of these hispines occur on *Heliconia.* Eggs are laid on leaves, stems or inflorescences, and larvae strip various surfaces of the plant. Pupae are hidden beneath dead leaves on the stalk or in wilted inflorescences (Strong, unpubl.).

Rolled leaves are natural sampling units for these hispines, and tests for interspecies displacement can be based upon leaf-occupancy patterns. Displacement is the most immediate and distinct result of interspecies competition, and is a cornerstone of evidence that competition is important among vertebrates, benthic organisms and plants. Displacement results in patterns of interspecies segregation. Examples for vertebrates are the interspecies exclusion of birds (Diamond 1978), chipmunks (Brown 1971), voles (Koplin and Hoffman 1968), kangaroo rats (Bowers and Brown, unpubl.) and other rodents (Grant 1972), and interspecific preemption of feeding sites by herbivorous fish on coral reefs (Sale and Dybdahl 1978). For intertidal invertebrates the crushing and crowding of barnacles (Connell 1961) and overgrowth of invertebrates and algae (Dayton 1975) are examples of rapid intragenerational displacement that yields monospecific patches of these organisms. Displacement among species of plants can be caused by shading or because moisture and nutrients are drawn from the soil more rapidly by one species (Harper 1977). Allelopathy is a mechanism of habitat fouling that can also displace and segregate plant species locally (Mueller 1966).

Displacement between hispine species could cause segregation among the leaves of a host population, analogous to the segregation of reef fish among feeding sites, of desert rodents among sand dunes, or of tree species among regeneration sites in a forest. Interspecies displacement among hispines could tend to segregate the higher densities of different species into different leaves of a host population.

For this report I would like to present part of a comparative and experimental analysis of interspecies segregation of hispines in leaf samples from Central America and Trinidad (Strong, unpubl.). The data are from 156 hispine populations from 53 communities. Seven of the communities contained only 1 hispine species, 12 had only 2 species, 13 had 3 species, 19 had 4 species, and 2 contained 5 species (Table 10-1). Most samples came from areas of no more than four hectares, from rolled leaves that

Table 10-1. Frequencies of rolled-leaf hispine species and their *Heliconia* hosts in the 53 community samples from Central America and Trinidad.

Host	Frequency	Hispine species	Frequency
H. wagneriana	6	*Cephaloleia vicina*	48
H. platystachys	4	*C. puncticollis*	37
H. curtispatha	1	*C. nigripicta*	9
H. vellirigera	2	*C. consanguinea*	9
H. imbricata	9	*C. instabilis*	8
H. latispatha	17	*C. neglecta*	4
H. mariae	5	*C. ornatrix*	4
H. tortuosa	2	*C. curtispatha*	3
H. wilsoni	7	*C. species novum*	1
		Chelobasis bicolor	14
		Ch. perplexa	11
		Nympharescus separatus	4
		Xenarescus monocerus	4

were within sight of another *Heliconia* plant of the same species. This excludes isolated plants from the samples. Rolled leaves on plants that were much more isolated than those included in the samples commonly contain densities of beetles as high as those in the samples. The beetles are strong flyers and find leaves well away from those that they have left, so my sampling criteria tend to produce homogeneous samples. I operationally define a hispine community as the beetles within the leaves of a single host species at one site at one time. This means that some of the communities that I sampled are subsets of larger communities that include more *Heliconia* and hispine species. The actual effect of this is small, because in most of my sampling areas over 90% of the rolled leaves were of one *Heliconia* species, and the rarer hosts do not provide refuge for hispines that might be displaced from the more locally abundant host species.

The analysis is based upon leaf co-occupancy of hispine species within communities. It asks if the higher density of each hispine species tends to occur apart, in a different subset of host leaves. This is a pattern predicted by simple interspecies displacement. The opposite of segregation is aggregation, and would be the pattern produced if the higher density of each species occurred in the same subset of leaves. The analysis can be envisioned by reference to one of the samples (Table 10-2). I have chosen a small sample for convenience, from Finca La Selva, Costa Rica, collected in August 1973. The host is *H. imbricata.* Table 10-2A gives the densities of the 4 species as they occurred in the 7 leaves of the sample. The columns are hispine species, the rows are leaves, and each entry is the number of beetles of a particular species in a particular leaf. Displacement might be inferred from the fact that the highest density of *Cephaloleia vicinia* (4 beetles, in leaf 7) occurs only with lower densities of *C. puncticollis* (0 beetles) and *C. consanguinea* (0 beetles). On the other hand, the opposite tendency might be inferred from leaf 2, which has the highest density of *C. consanguinea* (10 beetles) together with relatively high densities of all of the other 3 species.

Measuring the summed pattern among leaves would be no problem were only two hispine species in the sample; correlation measures give the degree of segregation or aggregation for a pair of species. But correlation coefficients among species pairs in samples with more than two species are not usually statistically independent, and cannot easily be interpreted together. For example, if the densities of the first and second and first and third species are both negatively correlated, the second and third species may be expected to be positively correlated. Averaging the coefficients might tend to result in values closer to zero than to any of the three.

The observed sum of row products for the sample in Table 10-2 equals 91. Because a row product with any zeros equals 0, some relatively high-density rows, such as 4 and 7, do not contribute at all to the sum of the row products. To remedy this, one can translate the matrix by adding 1 to each entry. This gives for an empty row (leaf) a value of 1, and each occupied leaf contributes to the sum of row products.

But what is a relatively high or low sum of row products for a sample? These can be calculated by a randomization or "Monte Carlo" procedure (Sokal and Rohlf 1969). What is needed is a frequency distribution of sums of row products for each sample, under the null hypothesis that the densities of each beetle species occur in leaves independently of the densities of other species. Frequency distributions of null sums of row products can be made by computer, by their repeated calculation for the matrix that is modified each time to randomize the leaf locations of beetle densities within

Table 10-2. Matrices of beetles in leaves. The observed sample (A) is from Finca La Selva, Costa Rica, and its host is *Heliconia imbricata*. Rows are the contents of individual leaves in the sample, columns are the hispines species. Matrix B is a rearrangement within columns (within species) of matrix A to show a high degree of interspecific segregation. Matrix C is a rearrangement within columns to show a high degree of interspecific aggregation. Row products for the untranslated matrices are under "+0", for the translated matrices under "+1"

Leaf	Cv	Cp	Cc	Chp	Row products (+0)	Row products (+1)
A: Observed						
1	1	1	1	1	1	16
2	3	3	10	1	90	352
3	0	1	2	1	0	12
4	0	3	0	1	0	8
5	0	0	3	0	0	4
6	0	1	0	1	0	4
7	4	0	0	0	0	5
Sum of row products					91	406
B: Rearranged for segregation						
1	0	0	10	0	0	11
2	0	0	3	1	0	8
3	0	1	2	1	0	12
4	3	1	1	1	3	32
5	4	1	0	1	0	20
6	1	3	0	1	0	16
7	0	3	0	1	0	8
Sum of row products					3	107
C: Rearranged for aggregation						
1	4	3	10	1	120	440
2	3	3	3	1	27	128
3	1	1	2	1	2	24
4	0	1	1	1	0	8
5	0	1	0	1	0	4
6	0	0	0	1	0	2
7	0	0	0	0	0	1
Sum of row products					149	607

Abbreviations: Cv, *Cephaloleia vicina*; Cp, *C. puncticollis*; Cc, *C. consanguinea*; Chp, *Chelobasis perplexa*.

species. Each random matrix is made by scrambling the observed densities of beetles within columns (within species) before multiplying within rows. The distribution of sums of row products of many matrices of this sort will give the range and frequency of possible sums of row products under the null hypothesis for that sample. The probability of a sum as high or low as that observed can be determined by comparison with the frequency distribution.

Two possible rearrangements of the example matrix of beetle species in leaves are shown in Tables 10-2B and C. Segregation among species is suggested by Table 10-2B: the higher densities of each species occur in different leaves, and the sum of row products is relatively small. Interspecies aggregation is suggested by Table 10-2C: the higher densities occur together in the same subset of leaves, and the sum of row products is relatively large.

A frequency distribution of sums of row products for 100 randomized or null matrices from the sample is shown in Fig. 10-1. The distribution for the actual densities is shown at the left, and that for the matrix translated by the addition of ones is shown at the right. The sum of row products of the actual, observed matrix is indicated by an arrow. For the untranslated matrix, the observed sum of row products falls above 95 of the 100 nulls. For the translated matrix the sum fall above 94 of the 100 nulls, indicating that this sample was relatively aggregated interspecifically. Were the observed values smaller than most null values, interspecific segregation would be indicated.

The validity of the technique can be tested with communities that have only two species, by comparison of the correlation coefficient with the proportion of nulls that exceed actual values. The relationship between Spearman's correlation coefficient and the proportion of nulls exceeding actual is shown in Fig. 10-2. As predicted, small proportions have positive correlation coefficients and large proportions have negative correlation coefficients. The community with a correlation coefficient of zero has approximately half of its nulls exceeding the actual (= 39). Darkened circles indicate significant correlation coefficients.

The values for the 46 multispecies community samples are shown relative to their null distributions in Fig. 10-3. Here the right end of the abscissa indicates interspecies displacement (a large fraction of nulls exceeding actual) and the left end indicates interspecies aggregation (a small fraction of nulls exceeding actual). Fig. 10-3 has one point for each community sample. The distribution of values in Fig. 10-3 does not indicate that interspecies competition by displacement among leaves is frequent in hispine communities, but rather that the opposite tendency is more common. Only 4 communities fall above the midpoint (50 nulls greater than actual) for the untranslated matrices, and only 13 for the translated matrices. The points tend to group at the low end of the graph. Some samples are actually at the lowest possible point, indicating that their communities are as aggregated within a subset of leaves as any combinations of the observed densities could be, by this method.

2 Leaf Homogeneity

An important consideration in analyses of association is the relative quality of sampling units (of the leaves within a sample in this case). Patterns of association can be affected by differences among the sampling units as well as by interactions among the organisms that occupy them. For hispines, rolled leaves of differing desirability or availability would tend to cause patterns independent of any interactions among the beetles. The pattern caused by any leaf heterogeneity within samples depends upon how beetles respond. Aggregation would be caused by uniform response of all beetle species to leaf heterogeneity, which would group all species into a subset of the leaves. This

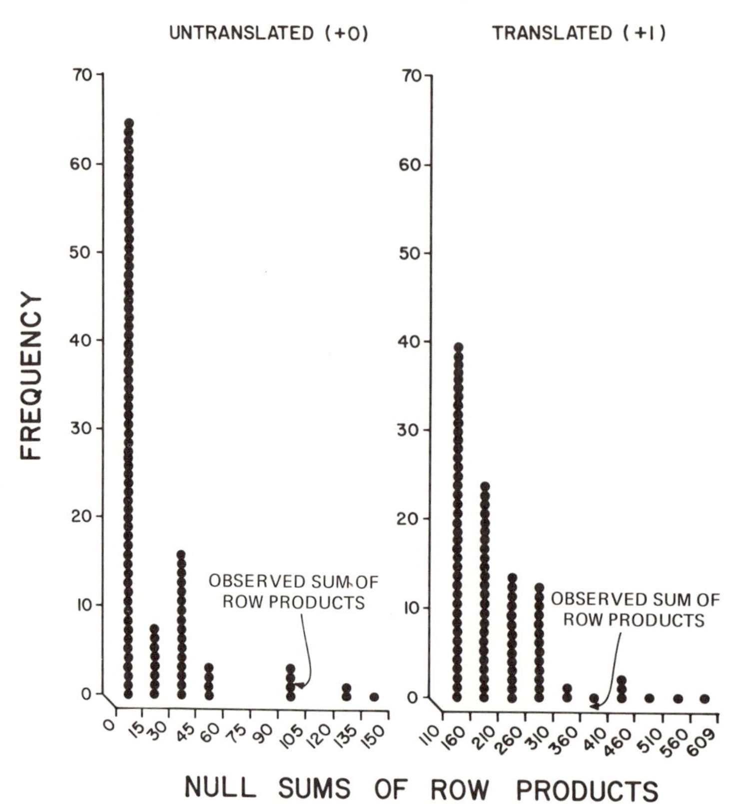

Figure 10-1. Frequency distributions of sums of row products for matrices which have been randomized within columns (within species). Each randomized matrix yields 1 point. The relative position of the sum of row products for the observed matrix is indicated by an arrow. Observed sums which fall to the right (high) in the null distribution indicate interspecies aggregation. Observed sums falling to the left indicate interspecies segregation. Translated matrices have the value of 1 added to each entry, in order that each row in the matrix contribute to the sum (see Table 10-2 and text).

would be identical to the effect caused by direct attraction among species independently of leaf heterogeneity.

The opposite effect, interspecies segreation, would be produced by leaf heterogeneity that beetle species respond to differently: for example, were one species to prefer leaves in the shade and another species to prefer leaves in the sun, the two would be segregated with the same pattern as would be produced by competition that causes interspecies displacement.

The tendency in the hispine communities is toward interspecies aggregation, so we need to examine the possibility of any leaf heterogeneity to which all species respond similarly. I have addressed this question experimentally with feeding tests on leaves found to contain differing densities of beetles (Strong, unpubl.). Briefly, the results are that beetles voraciously eat all leaves presented to them, and do not shun leaves that were found empty in nature. Also, I have compared hispine occupancy of *H. imbricata* and *H. latispatha* leaves over a complete range of microhabitats,

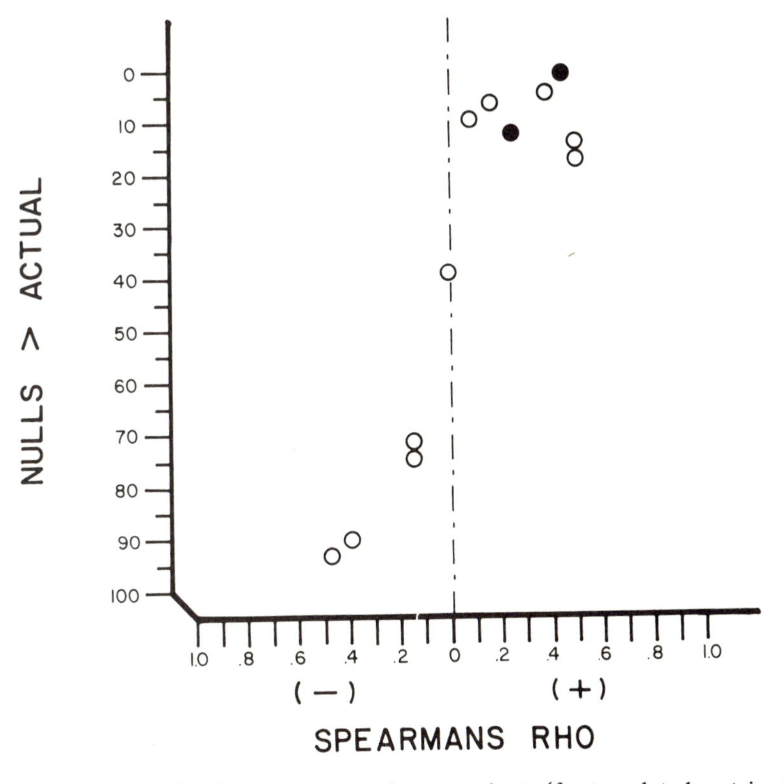

Figure 10-2. Relationship between sums of row products (for translated matrices) and Spearman's correlation coefficient for the communities with only two species. Open circles indicate nonsignificant correlation coefficients, closed circles significant coefficients.

and have found neither species composition nor density to be influenced by gradients of sun to shade, monospecies to multispecies host plant stands, or thick to open vegetation.

All 46 multispecies samples can be examined for patterns that might indicate leaf heterogeneity. One of the most obvious kinds of leaf heterogeneity would be a subset of undesirable or inaccessible leaves unwittingly included in the samples. These would be empty leaves. The problem is to calculate how many empty leaves are expected were species randomly associated among homogeneous leaves. The mere presence of empty leaves in a sample does not tell us much, as a number of these can be expected with neither interspecies attraction nor leaf heterogeneity. Leaves with no beetles are expected in community samples if all species individually occupy less than all of the leaves, which is common for hispines. Probabilities of numbers of empty leaves expected from random association of species among homogeneous leaves can be calculated using the observed numbers of occupied and vacant leaves for each individual species in a sample (Pielou and Pielou 1967).

Only 6 of the 46 multispecies communites have significantly more than the expected number of completely empty leaves (Strong, unpubl.). We can do more with these calculations, however, and ask if the statistical significances of empty leaf

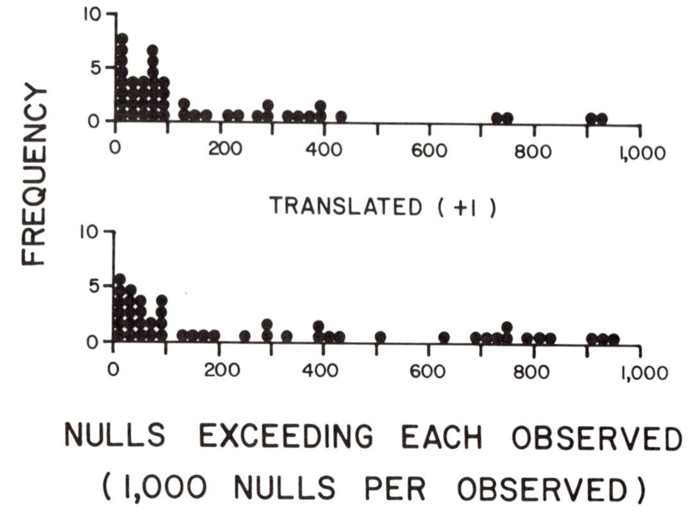

Figure 10-3. The position of the observed sum of row products in the distribution of null products, for each of the 46 multispecies hispine communities. Interspecies segregation is indicated by points falling toward the right end of the graph. Points toward the right have most nulls exceeding actual values. Interspecies segregation is indicated by points which fall toward the left end of the graph.

frequencies are correlated with the proportion of empty leaves in the sample. If empty leaves are generally undesirable or inaccessible to the hispines, we would expect the community samples with more empty leaves to be less likely to have that observed number on the basis of chance alone. Thirty-six of the multispecies community samples had empty leaves; samples without empty leaves have no possibility of an excess number of empty leaves. Among the 36 with empty leaves, no relationship occurs between the significance of frequency of empties and the proportion of empties ($r = 0.09$, n.s.). Less than 0.8% of the variation in the significances is explained by the proportion of empty leaves. The fraction of explained variation is not increased by use of arcsine-transformed proportions, because few of the observations were close to 0 or 1, where arcsine transformation would cause any great difference. From this we may conclude that the communities may have a weak or inconsistent tendency to have too many empty leaves (6 of the 46 communities had significantly more than expected), but that this tendency is not correlated with the number of empties observed. Thus, with this test we see no indication that empty leaves are generally undesirable. On the contrary, the test indicates that one expects a number of empty leaves in these communities, and that even the larger fractions of empties that one observes are consistent with random association of species among homogeneous leaves.

3 Factors Other Than Competition

The low densities of rolled-leaf hispines, very light grazing on the host plant, lack of agonistic interactions, and abundance of food and accommodation on their host *Heliconia* all suggest that neither intra- nor interspecies competition affects coexistence. What keeps beetle densities so low? Why aren't the plants so regularly defoliated that species are forced to compete for limited host space and food? The phytochemistry of *Heliconia* may be involved, but probably not in a density-dependent fashion. I suspect that parasitic hymenoptera are more important. Chemicals noxious to phytophages have not been found in the *Heliconia* so far investigated; no tannins, protein -complexing substances, enzyme inhibitors, alkaloids, saponins, cardiac glycosides or bufodieolides can be detected in leaf tissue eaten by the large number of herbivore species on *Heliconia* (Gage and Strong 1981). Nitrogen concentration in the tissues of the plant is relatively low (less than 2%), and some cells contain calcium oxylate crystals that may lacerate the gut of some herbivores, especially in early instars. Low nitrogen slows growth of lepidoptera on *Heliconia* but not of the hispines (Auerbach and Strong 1981). The primary mortality source of *Heliconia* herbivores is parasitic hymenoptera (Morrison and Strong 1981). For example, at La Selva, *C. consanguinea* loses up to 49% of its eggs to several species of eulophid wasps and to a species of *Trichogramma* (Morrison and Strong 1981). Pupae also suffer mortality from parasitic hymenoptera in excess of 30% (Strong, unpubl.).

4 Discussion

Community associations of adult *Heliconia* hispines are especially accessible to investigation because the beetles are concentrated in rolled leaves. Competition that led to displacement among species, such as that seen in plants, vertebrates and marine invertebrates, would cause rolled-leaf hispine species to be segregated among leaves. Interspecies segregation by displacement would be direct evidence for competition and is involved in the most convincing examples of how the phenomenon influences communities of organisms other than insects. The analysis presented here suggests that hispine species do not displace one another from rolled leaves. Rather, a weak opposite tendency of interspecies aggregation is in the data. The highest densities of different species occur together more frequently than would be expected from random association. An analysis of the frequency of empty leaves in communities suggest that the weak trend toward interspecies aggregation is not the result of distinct heterogeneity among leaves (which can cause patterns of aggregation independent of interspecies effects). Aggregation may be the result of interspecies attraction of beetles among virtually homogeneous leaves. The mechanisms and function of any interspecies attraction in hispines are not clear from what is now known. Experiments with the community on *H. imbricata* at La Selva show that the beetles tend to remain preferentially in leaves that contain other beetles of the same or different species. Just as in the patterns among communities, the tendencies of aggregation in the experiments are relatively slight, and reactions of individual beetles are quite variable. The beetles may be responding to feeding damage or to beetle odors that are similar among species. Rolled-leaf hispines have no agonistic behaviors toward conspecifics or other species. In petri plates they eat, copulate and behave harmoniously at densities much larger than occur in nature.

Both the experiments and biogeographic comparisons imply that food and accommodation within rolled leaves are well in excess of the levels that would adversely affect the beetles or limit species number in communities. In general, over 90% of mature *Heliconia* leaves at the sites sampled in this study bear damage made by the beetles, but the average amount of damage from beetles is less than 0.5% of the leaf area. Average amount of each leaf eaten by all phytophages is slightly greater than 1%. Abundant leaf tissue remains in the leaves with even highest beetle densities. Experiments show that this remaining tissue is eaten as readily by the beetles as is any tissue in the leaf.

Most larvae of rolled-leaf hispines are less dense than adults. They occur on the stem at the base of the plant, in rolled leaves, on the stalk at leaf basis, on unfurled leaves and in the inflorescences. Experiments that elevate larval densities show that abundant accommodation and food occur in the places in which the larvae are found. Larvae do not leave with higher frequency, or consume less per individual at experimentally elevated densities, for species that are convenient for experiments. The only species that reach high larval densities are the inflorescence-breeders (*Cephaloleia puncticollis* in Central America, *C. neglecta* in Venezuela and Trinidad). In both Central and South America, larvae of a second hispine species can sometimes be found in the inflorescences of some *Heliconia*: *Xenarescus monoceros* in the south and *Chelobasis perplexa, Ch. bicolor* and *Nympharescus separatus* in Costa Rica and Panama. The larvae of these "occasional-inflorescence" species normally feed in rolled leaves. They grow very slowly (Auerbach and Strong 1981) and use several rolled leaves for development, hiding on the stalk between successive rolled leaves. The inflorescence is the terminal meristematic activity of a *Heliconia* stalk. After flowering, stalks produce no more rolled leaves. The *Xenarescus, Chelobasis* and *Nympharescus* larvae that are found in inflorescences are individuals that have been stranded on the plant after the last rolled leaf. Larvae move only between the closest plants, and only if foliage of the two is touching. They cannot cross open ground. In Central America the larval density of the inflorescence-breeding *C. puncticollis* is not as high as in the south and, even late in the flowering season, stranded *Chelobasis* or *Nympharescus* larvae find much uneaten tissue in inflorescences. In Venezuela and Trinidad, *C. neglecta* reach much higher densities and can so reduce the tissue availability that *Xenarescus* larvae which have been stranded in the inflorescence have slightly reduced growth rates (Seifert and Seifert (1979). The effect on the *Xenarescus* population is probably very slight. Probably on a small fraction of *Xenarescus* larvae are exposed to this competition with *C. neglecta.* For example, in 1974, at the end of the dry season in Trinidad and Tobago, I found that fewer than 5% of the immature *Xenarescus* were pupae or larvae in inflorescences; the rest were in rolled leaves or on stalks that had not flowered. The inflorescence competition is not sufficient to kill larvae (Seifert and Seifert 1979) so this fraction would not have been greater were it not for the competition. The Seifert and Seifert (1979) discovery shows that competition among hispines can be detected when it occurs, and that it has no measurable effect on the community. *Xenarescus* are not less abundant in the south than are *Chelobasis* and *Nympharescus* in the north.

The results of this study are similar to several that deal with folivorous insects at higher latitudes. In these, proximal interspecies competition was not shown to structure communities, and densities were sufficiently low that plant resources were always abundant (Rathcke 1976, Broadhead and Wapshere 1966, Shapiro 1974). Other studies have found competition, but it is insufficiently intense or constant to cuase competitive exclusion (McClure and Price 1976, Addicott 1978, Denno 1980, Stiling 1980). These cases of competition involve similar species of homoptera (leafhoppers and aphids) that reach great densities within single growing seasons. Stiling (1980) has shown intraspecies competition to be more intense than that among species of leafhoppers of Welsh nettles. The obviously high densities in these cases and the relatively low densities in the cases where competition has been sought but not found suggest a pattern, but more critical experimental studies are needed to determine the frequency of competition in natural communities of phytophagous insects.

The results of studies with nonfolivorous insects are mixed. Granivorous ants can be sufficiently dense to deplete the seeds of desert soil and affect rodent populations deleteriously (Brown and Davidson 1977). But similar, coexisting species of ground beetles have been shown by long-term experiments to have no measurable effect upon one another (Wise 1981). Certainly, the few studies to date demonstrate great variation among communities in the occurrence of negative interactions among species at the same trophic level. Contrary to assumptions about the commonness of interspecies competition (Hutchinson 1977), quite similar species of insects coexist without negative influences upon one another, and species that do compete do not face inevitable competitive exclusion. Parasitism and host plant factors such as chemistry and phenology greatly affect phytophagous species, and impart much of what is interpreted as structure to communities in both temperate and tropical latitudes.

References

Addicott, J. F.: Niche relationships of aphids feeding on fireweed. Can. J. Zool. 56, 1837-1841 (1978).

Auerbach, M. A., Strong, D. R.: Nutritional ecology of *Heliconia* herbivores: experiments with plant fertilization and alternative hosts. Ecol. Monogr. (1981) (in press).

Bowers, M. A., Brown, J. H.: Body size and coexistence in desert rodents: chance or community structure? Unpubl. ms.

Broadhead, E., Wapshere, A. J.: *Mesopsocus* populations on larch in England–the distributions and dynamics of two closely related species of *Psocoptera* sharing the same food resource. Ecol. Monogr. 36, 327-388 (1966).

Brown, J. H.: Mechanisms of competitive exclusion between two species of chipmunks. Ecology 52, 305-311 (1971).

Brown, J. H., Davidson, D. W.: Competition between seed-eating rodents and ants in desert ecosystems. Science 196, 880-882 (1977).

Connell, J. H.: The influence of interspecific competition and other factors on the distribution of the barnacle *Chthamalus stellatus*. Ecology 42, 710-723 (1961).

Dayton, P. K.: Experimental evaluation of ecological dominance in a rocky intertidal algal community. Ecol. Monogr. 45, 137-159 (1975).

Denno, R. F.: Ecotope differentiation in a guild of sap-feeding insects on the salt marsh grass *Spartina patens.* Ecology 61, 702-714 (1980).

Diamond, J. M.: Niche shift and the rediscovery of interspecific competition. Am. Sci. 66, 322-331 (1978).

Gage, D. A., Strong, D. R.: The chemistry of *Heliconia imbricata* and *H. latispatha* and the slow growth of hispine beetle herbivore. Biochem. Syst. Ecol. (1981) (in press).

Grant, P. R.: Interspecific competition among rodents. Ann. Rev. Syst. Ecol. 3, 79-106 (1972).

Harper, J. L.: Population Biology of Plants. New York: Academic Press, 1977.

Hutchinson, G. E.: An Introduction to Population Ecology. New Haven-London: Yale Univ. Press, 1977.

Koplin, J. R., Hoffman, R. S.: Habitat overlap and competitive exclusion in voles (*Microtus*). Am. Midl. Nat. 80, 494-507 (1968).

MacArthur, R. H.: Geographic Ecology. New York: Harper and Row, 1972.

McClure, M. S., Price, P. W.: Ecotope characteristics of coexisting *Erythroneura* leafhoppers (Homoptera: Cicadellidae) on sycamore. Ecology 57, 298-940 (1976).

Morrison, G., Strong, D. R.: Spatial variations in egg density and the intensity of egg parasitism in a neotropical chrysomelid (*Cephaloleia consanguinea*). Ecol. Entomol. (1981) (in press).

Mueller, C. H.: The role of chemical inhibition (allelopathy) in vegetational composition. Bull. Torrey Bot. Club 93, 332-351 (1966).

Paine, R. T.: Intertidal community structure. Experimental studies on the relationship between a dominant competitor and its principal predator. Oecologia 15, 93-120 (1974).

Pielou, D. D., Pielou, E. C.: The detection of different degrees of coexistence. J. Theor. Biol. 16, 427-437 (a967).

Rathcke, B. J.: Competition and coexistence within a guild of herbivorous insects. Ecology 57, 76-87 (1976).

Sale, P. F., Dybdahl, R.: Determinants of community structure for coral reef fishes in isolated coral heads, lagoonal and reef slope sites. Oecologia 34, 57-74 (1978).

Seifert, R. P., Seifert, F. H.: Utilization of *Heliconia* (Musaceae) by the beetle *Xenarescus monoceros* (Oliver) (Chrysomelidae: Hispinae) in a Venezuelan forest. Biotropica 11, 51-59 (1979).

Shapiro, A. M.: Partitioning of resources among lupine-feeding lepidoptera. Am. Midl. Nat. 91, 243-248 (1974).

Sokal, R. R., Rohlf, F. J.: Biometry. San Francisco: Freeman, 1969.

Stiling, P. D.: Competition and coexistence among *Eupteryx* leafhopper (Hemiptera: Cicadellidae) occurring on stinging neetles (*Utrica dioica* L.). J. Anim. Ecol. 49, 793-805 (1980).

Strong, D. R.: Insect species richness: hispine beetles of *Heliconia latispatha.* Ecology 58, 573-582 (1977a).

Strong, D. R.: Rolled-leaf hispine beetles (Chrysomelidae) and their Zingiberales host plants in Middle America. Biotropica 9, 156-196 (1977b).

Strong, D. R.: Interspecific harmony among coexisting folivorous beetles. Unpubl. ms.

Wise, D. H.: Experimental evidence for the absence of interspecific competition in an insect community. Ecology (1981) (in press).

Chapter 11

What Makes a Good Island Colonist?

DANIEL SIMBERLOFF

1 Introduction

For one large community that I know well, I will discuss those aspects of the component species' life history or other ecological characteristics that determine whether they are good island colonists. Three broad questions will serve to focus the discussion:

(1) What traits and forces allow good colonists to be good colonists?
(2) Do any taxonomic or other patterns unify the set of good colonists?
(3) Are there rules or patterns describing which island colonists are found together?

The system that I will analyze is the animal community on red mangrove (*Rhizophera mangle*) islands of the Florida Keys, and most data are from my first experiment on this system, with E. O. Wilson (Wilson and Simberloff 1969, Simberloff and Wilson 1969, 1970, Simberloff 1969, 1976a, 1976b, 1978a, 1978b). Since this experiment is well known and the details described in cited works, I will not recapitulate them here, except to observe that six islands were defaunated by methyl bromide tent fumigation, then their subsequent recolonization by aboreal arthropods censused approximately every two weeks for a year and less frequently thereafter. I shall discuss here six tabulations of which species immigrated or became extinct when. For each of 236 species seen on these islands, there is a maximum of 92 two-week periods (divided among the six islands) during which a species could have immigrated or become extinct. Since a species can immigrate (by definition, Simberloff 1969) only if it is *not* already present, and can be extinguished only if it *is* already present, the total number of potential immigrations plus potential extinctions is 92.

For each species, one can estimate immigration rate or probability per 14 days as observed immigrations/possible immigrations, and extinction rate or probability per 14 days as observed extinctions/possible extinctions. These rates can be converted to approximate daily rates by simply dividing by 14, although these underestimate true daily rates, since an observed event could have happened any time in the interval (or events could have happened within the interval and never have been observed: e.g., an immigration followed by an extinction). By looking just at the end of the interval one can never observe two or more events in 14 days, which could in fact have occurred. So one tends to underestimate actual rates. Furthermore, simple division to get a daily rate leads also to underestimation. For example, if one observed 3 extinctions during 10 14-day intervals, one would calculate the daily rate as 3 extinctions/(14 days × 10) = 0.0214 extinctions/day. But if the daily rate were really 0.0214, the probability of extinction in 14 days would be $1 - (1 - 0.0214)^{14} = 0.2613$, whereas the observed 14-day probability was 0.3. For the data in this study, several sample calculations show the calculation underestimate generally to be 10-15%, and I compare rates to one another, not to an absolute figure, so I used the shortcut and divided by 14 to calculate daily rates.

2 What Makes Good Colonists Good Colonists?

Table 11-1 lists, by order, numbers of species of terrestrial arthropods (excluding mites) found on these six islands, plus numbers found on all mangrove islands that I have censused in the Keys (27) and, for selected groups, numbers found in all habitats of the Keys. One observes immediately that, of all Keys arthropods, fewer than 1/10 colonize mangrove islands. Of those found on mangrove islands, only 2/3 are routinely found on islands as small as these six, but a large fraction of these are very frequent colonists. If one arbitrarily defines a "good" colonist as one with a calculated daily extinction rate less than 0.0174 (the maximum possible, given the observation schedule) and daily immigration rate greater than 0.0016 (twice the minimum possible), one finds that about half the species observed on the six islands are good colonists. The remainder immigrate no more frequently than once a year, and are almost always quickly eliminated without breeding. Usually they do not breed because they are not adapted to the rather peculiar environment that constitutes a mangrove island, even if they occasionally disperse to them (Simberloff 1978b). They may not be arboreal, or may be host-specific for plants other than *Rhizophora,* etc.

One may now ask what traits or forces allow the good colonists to colonize, and distinguish two general, non-exclusive possibilities. A good colonist may have intrinsic biological characteristics that allow it to reach islands and suit it to the physical environment that it finds there. Or a good colonist may usually be a species that is able to insinuate itself into the existing community. I will treat the latter contention shortly, and for now concentrate on the former. At least during the first 8 months of this experiment, when population sizes were abnormally low or even zero, it is unlikely that species interacted much. Even in the complete absence of interactions an island community will nonetheless reach a species number equilibrium (Simberloff 1969) of:

$$\hat{S} = \sum_{\alpha=1}^{P} \frac{i_\alpha}{i_\alpha + e_\alpha}$$

where

P = size of species pool,
i_α = invasion rate of propagules of species α,
e_α = extinction rate for populations of species α.

Table 11-1. Breeding arthropod colonists of Florida Keys mangrove islands (mites excluded)

Group	Florida Keys	All mangrove islands	These six islands	Year 1 only	"Good" colonists	"Super" colonists
Silverfish		1	1	1	0	0
Collembola		5	2	1	0	0
Embioptera		3	1	1	0	0
Roaches		4	3	3	2	2
Crickets		8	5	5	4	4
Termites		3	2	2	0	0
Earwigs		1	1	1	1	0
Beetles		61	34	29	13	3
Thrips		11	7	6	5	1
Psocopterans		32	26	24	14	3
Homoptera		25	23	23	9	1
Neuroptera		4	3	3	3	2
Leipdoptera	555	15	13	13	10	3
Diptera		9	9	9	3	0
Wasps		66	44	42	21	5
Bees		2	1	1	1	0
Ants	50^+	18	17	17	10	9
Spiders		52	38	37	19	7
Pseudoscorpions		7	2	1	0	0
Other arthropods		7	4	0	0	0
Total	5000^+	334	236	219	115	40

One sees that two biological characteristics may render a species a good colonist: a high invasion rate and a low extinction rate. Either characteristic raises the value of that species' term in the above sum.

Figures 11-1 and 11-2 portray broad distributions of values for invasion and extinction rates, respectively. Mean daily extinction rate (0.0443) is about 16 times daily invasion rate (0.0027), which is why the number of species on any of these islands is just a small fraction of the size of the pool. Of course, most species in the pool do not appear at all in the above equation, because either their invasion rates are so low or extinction rates so high that their terms in the sum are nearly zero. Even of those that appear, about half (Fig. 11-2) are obligate transients, with extinction rates so high that they contribute almost nothing to the sum. These I have termed "noise" (Simberloff 1978b). In another experiment on mangrove island arthropods I observed a similar group of 96 species (of a 254 total) who were extinguished 111 times of 121 possible (Simberloff 1976c). If one inserts the estimated mean values of i_α and e_α for the non-noise species into the equilibrium sum, one finds S = 25 species. This is remarkably close to the observed equilibria, ca. 25-45 species (Simberloff and Wilson 1969, 1970), especially since for each species the islands would certainly have different i_α and e_α rather than mean i_α and e_α, and since the means themselves are just estimates for reasons stated earlier.

Species biologies are often viewed as compromises (e.g., Lewontin 1965, Pianka 1970, 1978), with components such as size and age of first reproduction each confer-

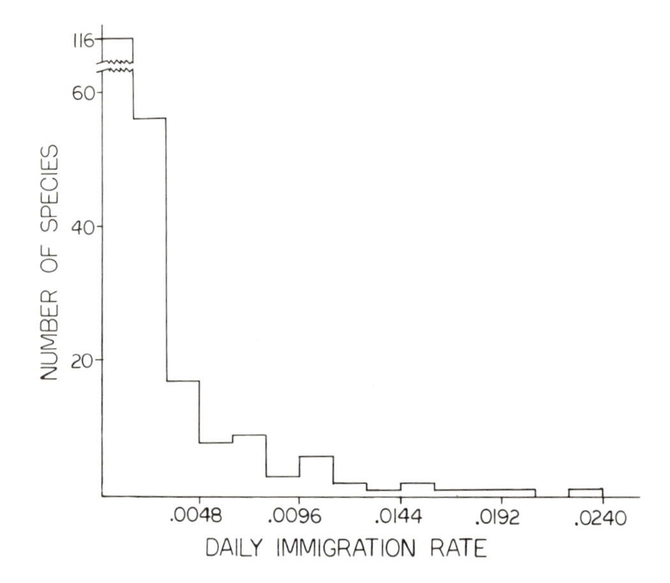

Figure 11-1. Distribution among mangrove colonists of daily immigration rates.

ring advantages or benefits while incurring disadvantages or costs. From this standpoint one might ask whether the physiological traits that make a species a high-frequency invader also tend to make it likely to become quickly extinct. Surprisingly, for the mangrove arthropod community precisely the opposite is true: invasion rates are *negatively* correlated with extinction rates at $r_{197} = -0.556, P \ll 0.01$. Good dispersers tend to be good persisters. If one arbitrarily defines the 27 species with the highest invasion rates ($i_\alpha > 0.0064$) plus the 27 species with the lowest extinction rates ($e < 0.01$) as "supercolonists,"[1] one finds strong confirmation of the tendency among good colonists toward a coordinated life history adapting them to island life (Table 11-2). Fourteen species (more than half of each life) of supercolonists are among both the top 27 dispersers *and* the top 27 persisters (Fig. 11-3). There are exceptions to the strong correlation between dispersal and persistence ability. For example, *Ariadna arthuri* (a nonballooning spider), *Lophoproctinus bartschi* (a polyxenid millipede) and *Rhyscotus* sp. (an oniscid isopod) are poor dispersers that rarely become extinct, while several bugs and spiders frequently invade mangrove islands but rarely persist (Simberloff 1976c). Among supercolonists, the ichneumonids *Calliephialtes ferrugineus* (No. 31) and *Casinaria texana* (No. 32) and the sphecid *Trypargilum johannis* (No. 33) similarly combine high invasion rates with mediocre persistence abilities. The overall correlation between invasion and extinction rates, however, is striking.

The apparent tendency towards a coordinated colonization "strategy" could be a statistical artifact. Brown and Kodric-Brown (1977) observed for insects colonizing individual flowers that a species' propagules may invade so frequently that one will never observe an extinction unless one censuses very often. They termed this the "rescue effect." If this were important, an apparently strong ability to persist would

[1]Using only the criterion of strong persistence, I delineated a group of 22 well-adapted colonists in another experiment (Simberloff 1976c); 12 are supercolonists.

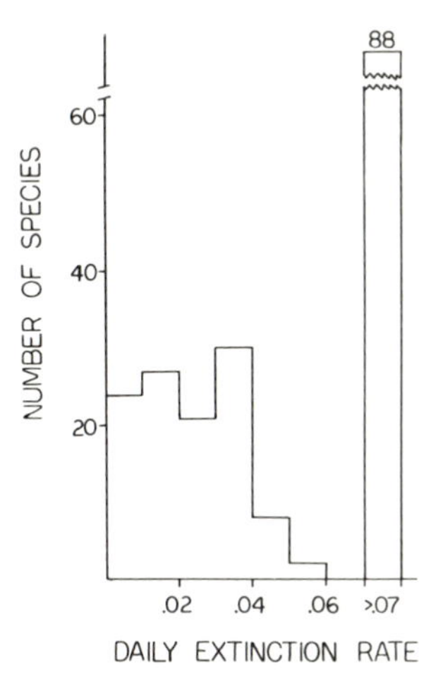

Figure 11-2. Distribution among mangrove colonists of daily extinction rates.

be generated *not* by longevity and reproduction, but rather by very frequent invasion; no breeding at all would be required. To ascertain whether such superdispersal could be the cause of many apparently super-low extinction rates, I used the circumstance that there were originally six sterile islands. For each species, then, I recorded the six time intervals required for the species to reach for the first time the six respective islands. The mean of these six values, or "mean interinvasion interval" (Simberloff 1976b), estimates the average time between successive arrivals of a species' propagules on an island as far from shore as these were (2-533 m). If a species never reached an island, I took the entire length of the experiment as the interval, so the estimate is very conservative. Clearly, mean interinvasion interval is just another statistic describing dispersal ability, and it is very strongly negatively correlated with the first such statistic, daily immigration rate: $r_{224} = -0.9161, P \ll 0.01$ (sample size is different for this correlation than for the correlation between invasion rates and extinction rates, since a number of species originally present never recolonized, so could not be tallied for the latter correlation).

The average (over all species) of the mean interinvasion interval is 173 days, or about 12.5 of the approximate 14-day periods between censuses. The distribution of mean intervals is highly skewed, with many more than half the species having mean intervals greater than this average. Since most species in this community have generation lengths of the order of 3-4 of these periods, they cannot be maintained primarily by repeated invasion rather than breeding. Those supercolonists with the 27 lowest extinction rates also have lower than average mean interinvasion intervals: over half are in the bottom 20%. But even these usually have mean interinvasion intervals too long for sustained existence simply by virtue of continuous invasion: their average mean interval is 116 days, or 8.25 14-day periods. So they must, on average, persist over 16 weeks without benefit of invasion. To sum up, the rescue effect may contribute to, but cannot by itself account for, the existence of some particularly persistent island colonists.

Table 11-2. The "supercolonists" (numbers as in Fig. 11-3)

1.	*Latiblattella n. sp.*	(Blattidae)
2.	*Latiblattella rehni*	(Blattidae)
3.	*Cyrtoxipha confusa*	(Gryllidae)
4.	*Orocharis gryllodes*	(Gryllidae)
5.	*Tafalisca lurida*	(Gryllidae)
6.	*Cycloptilum spectabile*	(Gryllidae)
7.	*Tricorynus* sp.	(Anobiidae)
8.	*Styloleptus biustus*	(Cerambycidae)
9.	*Pseudoacalles sablensis*	(Curculionidae)
10.	*Pseudothrips inequalis*	(Thripidae)
11.	*Caecilius incoloratus*	(Caeciliidae)
12.	*Peripsocus pauliani*	(Peripsocidae)
13.	*Psocidus texanus*	(Psocidae)
14.	*Aleurothrixus* sp.	(Aleyrodidae)
15.	*Chrysopa collaris*	(Chrysopidae)
16.	*Chrysopa rufilabris*	(Chrysopidae)
17.	*Phocides pygmalion*	(Hesperiidae)
18.	*Ecdytolopha* sp.	(Olethreutidae)
19.	*Bema ydda*	(Phycitidae)
20.	*Iphiaulax epicus*	(Braconidae)
21.	*Pachodynerus nasidens*	(Eumenidae)
22.	*Camponotus abdominalis*	(Formicidae)
23.	*Crematogaster ashmeadi*	(Formicidae)
24.	*Paratrechina boubonica*	(Formicidae)
25.	*Pseudomyrmex elongatus*	(Formicidae)
26.	*Pseudomyrmex "flavidula"*	(Formicidae)
27.	*Xenomyrmex floridanus*	(Formicidae)
28.	*Paracryptocerus varians*	(Formicidae)
29.	*Monomorium floricola*	(Formicidae)
30.	*Tapinoma littorale*	(Formicidae)
31.	*Calliephialtes ferrugineus*	(Ichneumonidae)
32.	*Casinaria texana*	(Ichneumonidae)
33.	*Trypoxylon johannis*	(Sphecidae)
34.	*Eustala* sp. 1	(Araneidae)
35.	*Gasteracantha ellipsoides*	(Araneidae)
36.	*Ayscha velox*	(Clubionidae)
37.	*Sergiolus* sp.	(Gnaphosidae)
38.	*Hentzia grenada*	(Salticidae)
39.	*Leucauge venusta*	(Tetragnathidae)
40.	*Tetragnatha antillana*	(Tetragnathidae)

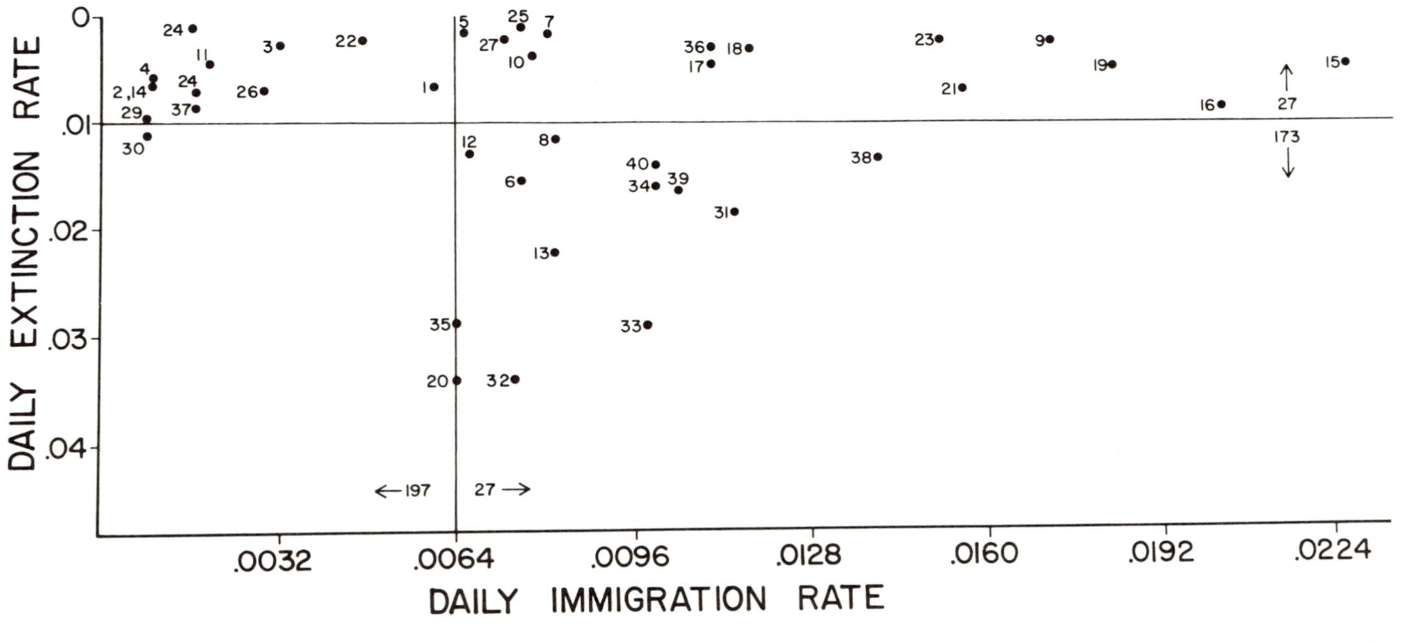

Figure 11-3. Daily extinction rate plotted against daily immigration rate for supercolonists. Vertical line divides the 197 poor dispersers from the 27 good dispersers; horizontal line divides the 173 poor persisters from the 27 good persisters (totals differ because extinction rate estimates not possible for some species). Note that species in upper right-hand corner are good at both dispersing and persisting. Most species in pool fall in lower left-hand corner. Roaches = 1, 2; crickets = 3-6; beetles = 7-9; thrips = 10; psocopterans = 11-13; whitefly = 14; green lacewings = 15, 16; lepidopterans = 17-19; wasps = 20, 21, 31-33; ants = 22-30; spiders = 34-40. Species' names are in Table 11-2.

3 Patterns Among Good Colonists

There is no obvious taxonomic or other pattern to the set of 40 supercolonists (Fig. 11-3 caption); there is more than one way to skin a cat. Nine orders and 24 families are represented, and 14 species manifest the coordinated colonization strategy; 13 have only high invasion rates and another 13 have only low extinction rates. Nine ants comprise the largest single taxonomic groups of supercolonists; all have very low extinction rates (actually, colony extinction rates; we may well have missed extinction of founding queens who never produced workers). The 7 spiders are all supercolonists solely by virtue of their high invasion rates. Perhaps the best colonists of all in terms of both getting to the islands and staying there are the pair of green lacewings. Another interesting fact is that not all *Rhizophora* specialists are supercolonists. For example, a host-specific skipper (*Phocides pygmalion*) and two moths (*Ecdytolopha* sp. and *Bema ydda*) are supercolonists, but the host-specific moths *Alarodia slossoniae* and *Oxydia cubana* are not. The latter two perform somewhat better on larger islands (e.g., Simberloff 1976c) but, on the small islands that abound in the Keys, they are not among the best colonists.

The supercolonists do become extinct occasionally, even after they have built up sizeable populations; this is a manifestation of the dynamic nature of nearshore island communities (cf. Simberloff 1976b, 1976c). Supercolonists produced 141 extinctions in this experiment, and many more in others. No species is immune to this risk, though some may have very low extinction probabilities, and all have lower extinction probabilities on larger islands (Simberloff 1976c).

4 Which Colonists are Found Together, and Why?

Finally, one must address the question of whether rules or patterns constrain which species colonize islands together. A plethora of papers for vertebrates (e.g., Grant 1970, Diamond 1975, Abbott et al. 1977, King and Moors 1979) and one for plants (Abbott 1977) purport to show that species found together on islands are highly circumscribed by interactions among species, especially competitive interactions that preclude certain species from coexisting. By contrast, Pielou and Pielou (1968) suggested that those arthropods found in a series of bracket fungi are consistent, for the most part, with the hypothesis that species colonize brackets independently of one another. Although a detailed description of the statistical methods is beyond the scope of this symposium, Connor and Simberloff (1979) and Simberloff and Connor (1981) have shown that every study except the Pielous' (1968) is so beset by statistical and/or logical errors that the claims are untenable from the data presented. They have concluded that the Pielous' (1968) assertion (that a model of independent colonization is largely consistent with observed island biogeographic patterns) is valid for all studies claiming a major role for competitive exclusion in dictating which species colonize together. In a similar vein, Strong et al. (1979) and Simberloff and Boecklen (1981) have criticized on both logical and statistical grounds the voluminous literature claiming that morphological size relationships determine which species can coexist, and have shown that in few reported communities are species' sizes statistically different from those predicted by a model of independent, random, uniformly distributed sizes. None of this work proves that competitive interactions are *not* common and/or important in determining which species coexist; it demonstrates only that the biogeographic and size patterns do not show that they *are* common and/or important.

For mangrove island colonization, Heatwole and Levins (1972) claimed that, in addition to a species number equilibrium, a trophic structure equilibrium obtains, and that interspecific interactions constrain sets of species on an island to have given proportions of predators, parasitoids, wood-borers, etc. They arrived at this conclusion by classifying all colonists into eight trophic categories and noting how this octopartite distribution shifted as colonization proceeded. However, their observation that the defaunated islands' distributions became progressively more similar to one another (and to the predefaunation distributions) was an artifact of the sample-size dependence of the statistic with which they characterized differences between distributions (Simberloff 1976a). In fact, *randomly* drawn subsets from the pool of available Keys arthropods showed exactly the same trend for this statistic that Heatwole and Levins (1972) detected for the observed mangrove island communities. This is not to say that the communities *are* primarily random assemblages consisting of species that colonize independently of one another—only that the data have not yet been shown inconsistent with this hypothesis.

Recently I showed by Monte Carlo simulation that if one examines simply species' names, rather than general trophic characteristics, there is again little evidence that sets of species coexisting on mangrove islands are other than random, independent draws from the available pool (Simberloff 1978a). I constructed a species pool of all mangrove arthropods, weighted by their frequencies on 15 small islands, and compared actual number of species shared between the original community on any island and the community during recolonization with the number of shared species expected if the communities during recolonization were just random, independent draws. This procedure corrects for increasing community size during much of colonization. That observed numbers of shared species did not exceed expected by much between most pairs of communities, and that this excess did not increase as colonization proceeded, does *not* prove that species interactions do not constrain island colonization. For one thing, the species frequencies themselves on the 15 islands (used to derive the weights for the species pool) might have been strongly influenced by interactions. But my point is *not* that interspecific interactions do not exist; it is only that biogeographic patterns alone do not force us to conclude that they do, and are not very different from those that a simple model of random, independent colonization predicts.

Of course, species sets on islands are not just random draws, nor are all species unaffected by which other species are presented. For mangrove islands I have described a few instances in which the geographic patterns alone *imply* (do not prove) that certain species affect other species' local survival (Simberloff 1976c). But stylized interactions between specific pairs of species cannot be demonstrated simply by statistically examining large species lists; these lists must be affected by a welter of factors that submerge and confound the effects of interactions. Rather, experiment and laborious autecological research will almost certainly be required to demonstrate the importance of species interactions, not only on islands but in general.

Summary. Broadly speaking, a species may be a good island colonist by being very adept at reaching islands and/or very good at persisting on them once there. Good persistence, in turn, may derive from ability to tolerate either the physical environment of the island and/or the other species present. Mangrove island arthropods have broad distributions of both dispersal and persistence ability. Most species do not either reach islands frequently or survive on them for long. Of those that occur often, some are good dispersers, some are good persisters, and a surprising number are both: invasion and extinction rates are very highly negatively correlated among all species in the pool. The correlation is real, and not just an artifact of very frequent invasion's obscuring extinction when censusing is infrequent. Not only are good colonists "good" for different biological reasons, but strong taxonomic patterns are lacking.

As to whether a species' persistence ability on islands is heavily influenced by its interactions with other species, evidence to date is inconclusive, and not only for mangrove arthropods: most studies have been severely flawed logically and/or statistically, so the nearly universal conclusion that interactions (particularly interspecific competition) are crucial is premature. For the mangrove arthropods as for other communities there is as yet little indication that species do not colonize islands primarily independently of one another. This is not to claim that island colonization *is* a random, independent process—only that simple statistics based on large species lists are unlikely to be able to show that it is *not* random and independent, and will surely be unable to specify the nature of biotic interactions. Experiment and intensive autecological research are necessary.

Acknowledgments. I thank the convenors of this symposium, Drs. Robert Denno and Hugh Dingle, for inviting me to participate, and Dr. Elizabeth P. Lacey for assistance with the figures.

References

Abbott, I.: Species richness, turnover, and equilibrium in insular floras near Perth, Western Australia. Aust. J. Bot. 25, 193-208 (1977).

Abbott, I., Abbott, L. K., Grant, P. R.: Comparative ecology of Galapagos ground finches (*Geospiza* Gould): Evaluation of the importance of floristic diversity and interspecific competition. Ecol. Monogr. 47, 151-184 (1977).

Brown, J. H., Kodric-Brown, A.: Turnover rates in insular biogeography: Effects of immigration on extinction. Ecology 58, 445-449 (1977).

Connor, E. F., Simberloff, D.: The assembly of species communities: Chance or competition? Ecology 60, 1132-1140 (1979).

Diamond, J. M.: Assembly of species communities. In: Ecology and Evolution of Communities. Cody, M. L., Diamond, J. M. (eds.). Cambridge, Mass.: Harvard Univ. Press, 1975, pp. 342-444.

Grant, P. R.: Colonization of islands by ecologically dissimilar species of mammals. Can. J. Zool. 48, 545-553 (1970).

Heatwole, H., Levins, R.: Trophic structure stability and faunal change during recolonization. Ecology 53, 531-534 (1972).

King, C. M., Moors, J. P.: On co-existence, foraging strategy and the biogeography of weasels and stoats (*Mustela nivalis* and *M. erminea*) in Britain. Oecologia 39, 129-150 (1979).

Lewontin, R. C.: Selection for colonizing ability. In: The Genetics of Colonizing Species. Baker, H. G., Stebbins, G. L. (eds.). New York: Academic Press, 1965, pp. 77-94.

Pianka, E. R.: On *r* and *K* selection. Am. Nat. 104, 592-597 (1970).

Pianka, E. R.: Evolutionary Ecology, 2nd ed. New York: Harper and Row, 1978.

Pielou, D. P., Pielou, E. C.: Association among species of infrequent occurrence: The insect and spider fauna of *Polyporus betulinus* (Bulliard) Fries. J. Theor. Biol. 21, 202-216 (1968).

Simberloff, D.: Experimental zoogeography of islands: A model for insular colonization. Ecology 50, 296-314 (1969).

Simberloff, D.: Trophic structure determination and equilibrium in an arthropod community. Ecology 57, 395-398 (1976a).

Simberloff, D.: Species turnover and equilibrium island biogeography. Science 194, 572-578 (1976b).

Simberloff, Experimental zoogeography of islands: Effects of island size. Ecology 57, 629-648 (1976c).

Simberloff, D.: Using island biogeographic distributions to determine if colonization is stochastic. Am. Nat. 112, 713-726 (1978a).

Simberloff, D. S.: Colonisation of islands by insects: Immigration, extinction, and diversity. In: Diversity of Insect Faunas, Mound, L. A., Waloff, N. (eds.). Oxford: Blackwell Scientific Publ., 1978b, pp. 139-153.

Simberloff, D., Boecklen, W.: Santa Rosalia reconsidered: Size ratios and competition. Evolution (1981) (in press).

Simberloff, D., Connor, E. F.: Missing species combinations. Am. Nat. (1981) (in press).

Simberloff, D., Wilson, E. O.: Experimental zoogeography of islands: The colonization of empty islands. Ecology 50, 278-296 (1969).

Simberloff, D., Wilson, E. O.: Experimental zoogeography of islands: A two-year record of colonization. Ecology 51, 934-937 (1970).

Strong, D. R., Jr., Szyska, L. A., Simberloff, D.: Tests of community-wide character displacement against null hypotheses. Evolution 33, 897-913 (1979).

Wilson, E. O., Simberloff, D.: Experimental zoogeography of islands: Defaunation and monitoring techniques. Ecology 50, 267-278 (1969).

Chapter 12

Biogeographic Patterns among Milkweed Insects

RICHARD P. SEIFERT

1 Introduction

Biogeographers have recognized two main patterns associated with the distributions of organisms. First, it has commonly been reported that tropical environments have a higher species richness and diversity than do temperate environments (Darwin 1859, Wallace 1876, Richards 1952, MacArthur 1965, 1969, 1972, Pianka 1966, 1978, Slobodkin and Sanders 1969). Second, islands have a consistently depauperate fauna and flora when compared with mainland regions of similar areas and climates (MacArthur and Wilson, 1963, 1967, Allan et al. 1973, Carlquist 1974, Simberloff 1974). Janzen (1968, 1973a) has suggested that many of the patterns of species accumulation on islands should occur, both evolutionarily and ecologically, for the accumulation of insect species on their host plants. Studies of host plants as ecological islands for their insect fauna have included examinations of the relationship of host species geographic distribution to insect species richness (Opler 1974, Strong 1974a, 1974b, Strong et al. 1977, Cornell and Washburn 1979, Lawton and Price 1979) and the relationship of local host plant patch size to insect species richness (Seifert 1975, Raupp and Denno 1979, Denno et al., Chapter 9).

In this study I consider milkweed plants (Asclepiadaceae) as ecological islands for insects at four different locations including two tropical mainland sites, a tropical island site and a temperate mainland site. I intended to determine not only the extent to which island biogeography theory (MacArthur and Wilson 1967) could be applied to these ecological islands but also the extent to which patterns of tropical, temperate and island insect species richness caused variation in the ecological island patterns from the four sites. The plants which I considered as ecological islands come from a family of plants known to harbor cardiac glycosides and to have a specialized herbivorous insect fauna capable of incorporating the toxins into their fat bodies (Brower and Brower 1964, Brower 1969, Price 1975). These plants should have a lower number of transient, generalized herbivores due to the toxic nature of the secondary plant compounds. All of the data reported on in this paper came from times of high insect abundances, during the rainy season for the tropical sites and during the late summer for the temperate site.

2 Materials and Methods

Four field sites, each with a separate resident milkweed population, were chosen for this research. The first tropical mainland field site was located in a region of tropical, very dry forest, at an elevation of about 150 m, about 15 km east of Cata along the road between Cata and Cuyagua, Aragua, Venezuela (10°24'N, 67°20'W). (All records of forest type refer to the Holdridge (1964) life zone classification, which is based on climatic data and does not necessarily represent the extant local vegetation.) A population of *Calotropis procera* (W. Aiton) W. T. Aiton grows in an area of about 80 m^2 at this location. Records of the numbers of individuals and numbers of species of insects on individual plants were taken on 2, 3, 4 and 11 August 1977.

The second tropical mainland site was in a region of tropical dry forest, at an elevation of about 50 m, 4 km north of Ocumare de la Costa along the road between Ocumare and Cata (10°18'N, 67°30'W). In this location a population of *Calotropis procera* grows in a pasture of about 100 m^2. Data were collected on 8 and 13 August 1979.

The tropical island field site was located in a region of tropical dry forest 1 km north of Bouillante, at an elevation of 25 m, on the Caribbean island of Guadeloupe (16°15'N, 61°45'W). A population of *Calotropis procera* grows there in a pasture over an area of about 100 m^2. All field records from Guadeloupe came from 21 July 1978.

The temperate mainland field site was located along the edge of the road at the north entrance to the Capital Centre, Landover, Maryland (38°55'N, 76°45'W). There a population of *Asclepias syriaca* L. grows along a narrow strip about 10 × 80 m along the road edge entering the Capital Centre. Field collections from Maryland were made on 9 August 1978.

At this juncture I indicate two procedural complexities in my study. First, *Calotropis procera* is an introduced species in the Neotropics and therefore could have reduced insect species richness, simply because there has not been time for an herbivorous insect fauna to evolve. However, Strong et al. (1977) found that date of introduction had nothing to do with the number of insect pests on sugar cane: herbivorous insect species richness was dependent on the area planted in sugar cane. Thus, it is reasonable to assume that *C. procera* does not have a reduced complement of insects due to its recent introduction into the Neotropics.

Second, since each collection was made over a period of only a few days, some insect species associated with particular age classes of milkweeds were not represented in the sample. For example, the Capital Centre collection did not include any lygaeid bugs (Table 12-4), even though I have seen these bugs at Capital Centre on other dates. However, my study was designed to measure species richness on milkweeds at a particular (short) period of time, and was not meant to produce a complete list of Latin binomens associated with each milkweed population.

Data were collected similarly at each field site. The number of leaves on each plant were counted, the distance to the nearest individual plant was measured, and records were taken of the number of individuals of each insect species on each plant. Insects were separated into different species on morphological bases and, where possible, were identified to family. The insects on the milkweeds consisted primarily of herbivores but included some transient species (mostly flies), a few predatory species (predatory coccinellid beetles, nabid bugs and mantids in Venezuela, a syrphid fly larva feeding on aphids in Guadeloupe) and, when flowers were present, included some nectivorous or pollen feeding butterflies, flies, bees and beetles. Inflorescences occurred only on the largest milkweed plants at each of the sites.

Seven different biogeographic patterns were studied for the insects associated with the four milkweed populations. First, the faunal coefficient (MacArthur and Wilson 1963), z, was generated from the linear equation

$$\log(S+1) = \log C + z \cdot \log(A+1) \tag{1}$$

where S is equal to the number of insect species on each plant and A is the area of the plant as estimated by the number of leaves on the plant. (The constant of 1 was added to each side of the equation to eliminate situations where S = 0 and log S was undefined.) A linear regression of log (S+1) on log (A+1) was run for each milkweed population, and the slopes of the regressions (z) for all four sites were compared using an analysis of covariance. The faunal coefficient indicates the rate at which host plant area accumulates insect species. Second, for each milkweed population the number of insect species on each plant was regressed against the distance separating each plant from its nearest neighbor. This analysis was used to indicate whether distance effects, such as those shown to be important for true islands (MacArthur and Wilson 1967), are important in determining species richness for these ecological islands. Third, three Kruskal-Wallis one-way analyses of variance were run on the number of insect species per milkweed plant in each of the milkweed populations, in order to determine if the number of insect species per plant differed. These tests included an overall comparison of all four populations of milkweeds, as well as comparisons of Cata milkweeds versus Maryland milkweeds (tropical mainland vs. temperate mainland). Fourth, an analysis of variance on the square root transformation of number of insect species per leaf at each location was run. Fifth, a Kruskal-Wallis test was run to detect differences in the number of individuals per insect species at each location. Sixth, insect diversity, H, was computed for each plant using the Brillouin diversity index

$$H = (1/N) \cdot (\log N! - \sum_{i=1}^{s} \log n_i!) \tag{2}$$

where N is the total number of individuals of all insects on each plant, n_i the number of individuals of the ith species and s the total number of insect species (Pielou 1974). (The Brillouin diversity index was used in this study rather than the more commonly used Shannon-Wiener diversity index, since each insect on each plant was recorded.) Three Kruskal-Wallis analyses of variance were run to distinguish differences in insect diversity among the four milkweed populations. In all cases above where the Kruskal-Wallis tests were run, they were chosen after square root and logarithmic transformations applied to the data did not eliminate heteroscedacticity. Each Kruskal-Wallis test was run at the conservative probability level $P \leqslant 0.025$ to give an overall probability level for each series of three tests of $P \leqslant 0.05$. Finally, the evenness component of species diversity, J, for each plant was estimated from the formula

$$J = H/H_{max} = H/\{(1/N) \cdot \log N! - ((s-r) \cdot \log[N/s]! + r \cdot \log([N/s]+1)!)\} \quad (3)$$

where [N/s] is the integer part of N/s and r is N-s•[N/s] (Pielou 1974). Evenness was determined only for those milkweed plants which had two or more insect species. The evenness component cannot be evaluated mathematically when less than two species exist and, under such a situation, the evenness component has no biological meaning. A linear regression of evenness on area was run for each milkweed population to determine whether area influenced evenness.

3 Results

Tables 12-1-4 list the total number of insect species, determined to family when possible (following Imms 1957), found on milkweeds from Cata, Ocumare, Guadeloupe and Maryland, respectively. While many of these insects are host-specific, some of the rare insect species represent either transient species, predatory species or species attracted by nectar (when the plants were in flower).

For all milkweed populations there was a significant effect of area on species richness (F_s = 217.800, $P < 0.001$ for Cata; F_s = 16.206, $P < 0.001$ for Ocumare; F_s =

Table 12-1. Orders, families and number of insect species found on milkweeds from Cata, Venezuela

Order and family	No. of species	Order and family	No. of species
Orthoptera		Diptera	
Gryllidae	1	Chironomidae	1
Tettigoniidae	1	Otitidae	1
Acrididae	2	Muscidae	1
Hemiptera		Hymenoptera	
Homoptera		Mutillidae	1
Cicadellidae	3	Formicidae	4
Delphacidae	2	Vespidae	1
Fulgoridae	1	Apidae	3
Aphidae	2	Undetermined parasitic wasps	4
Coccidae	2		
Heteroptera		Coleoptera	
Miridae	1	Cantharidae	1
Lygaeidae	3	Nitidulidae	1
		Erotylidae (?)	1
Thysanoptera		Coccinellidae	1
Undetermined	1	Tenebrionidae	1
		Cerambycidae	1
Lepidoptera		Chrysomelidae	10
Danaidae	1	Curculionidae	1
Lycaenidae	2	Undetermined	1
Undetermined	1		

71.063, $P < 0.001$ for Guadeloupe; $F_s = 77.967, P < 0.001$ for Maryland) (Fig. 12-1-4). The faunal coefficients were quite different for the four milkweed populations. The Maryland population had the highest faunal coefficient (1.049), the Ocumare population had the lowest faunal coefficient (0.4735), while the faunal coefficient of the Cata population (0.7174) and the Guadeloupe population (0.5980) were intermediate. An analysis of covariance indicated that these four faunal coefficients were significantly different ($F_s = 4.4062$, $0.01 > P > 0.001$). The Maryland population appears to have the highest slope, because very small plants have no insects while larger plants have a high accumulation of insect species. Guadeloupe has a low slope because fewer potential colonizers exist for the plants, so that small plants may have a relatively high proportion of the available colonizers compared with the Maryland milkweed population. The low z value for the Ocumare population seems to be due to the relatively high proportion of all insect species there which are found on small- and intermediate-sized milkweeds. For the Cata milkweed population, small plants accumulate species from a larger species pool, producing a slope intermediate between those of Maryland and Guadeloupe populations.

Table 12-2. Orders, families and number of insect species found on milkweeds from Ocumare, Venezuela

Order and family	No. of species	Order and family	No. of species
Collembola		Thysanoptera	
Entomobryidae	2	Undetermined	1
Isotomidae	1		
		Lepidoptera	
Odonata		Danaidae	1
Libellulidae	1		
		Diptera	
Orthoptera		Chironomidae	2
Tettigoniidae	3	Stratiomyidae	1
Acrididae	2	Dolichopodidae	1
		Muscidae	2
Dictyoptera		Undetermined	2
Mantidae	2		
		Hymenoptera	
Hemiptera		Formicidae	5
Homoptera		Vespidae	1
Cicadellidae	2	Apidae	1
Fulgoridae	1	Undetermined parasitic wasps	4
Aphididae	3		
Coccidae	1	Coleoptera	
Heteroptera		Carabidae	1
Nabididae	2	Erotylidae (?)	1
Lygaeidae	1	Coccinellidae	2
Coreidae	2	Tenebrionidae	1
Tingidae	1	Chrysomelidae	11
Pentatomidae	1	Curculionidae	1
		Undetermined	4

Table 12-3. Orders, families and number of insect species found on milkweeds from Bouillante, Guadeloupe

Order and family	No. of species	Order and family	No. of species
Collembola		Syrphidae	1
Sminthuridae	1	Otitidae	2
		Drosophilidae	1
Hemiptera		Muscidae	1
Homoptera		Undetermined	2
Cicadellidae	2		
Aphidae	2	Hymenoptera	
Coccidae	1	Formicidae	5
Heteroptera		Apidae	1
Lygaeidae	1		
		Coleoptera	
Diptera		Lampyridae	1
Chironomidae	1	Coccinellidae	1
Dolichopodidae	1	Chrysomelidae	1

There is no statistically significant effect of distance on insect species richness for any of the four milkweed populations (Cata: $F_s = 0.8346, 0.50 > P > 0.25$; Ocumare: $F_s = 0.6965, 0.50 > P > 0.25$; Guadeloupe: $F_s = 0.0062, P > 0.75$: Maryland: $F_s = 0.2679, 0.75 > P > 0.50$). The number of insects on milkweed plants is not dependent on the distances among the plants.

There is a statistically significant difference among the number of insect species per milkweed plant from the four locations (Table 12-5). Not only is there an overall difference of insect species richness among the locations ($H = 36.3361, P < 0.001$),

Table 12-4. Orders, families and number of insect species found on milkweeds from Capital Centre, Maryland

Order and family	No. of species	Order and family	No. of species
Orthoptera		Dolichopodidae	3
Acrididae	1	Drosophilidae	1
Hemiptera		Lepidoptera	
Homoptera		Arctiidae	2
Membracidae	2		
Cicadellidae	3	Hymenoptera	
Aphidae	1	Formicidae	3
Coccidae	1		
Heteroptera		Coleoptera	
Nabididae	1	Scarabaeidae	1
Pentatomidae	1	Coccinellidae	3
		Cerambycidae	1
Diptera		Chrysomelidae	2
Chironomidae	1	Curculionidae	4

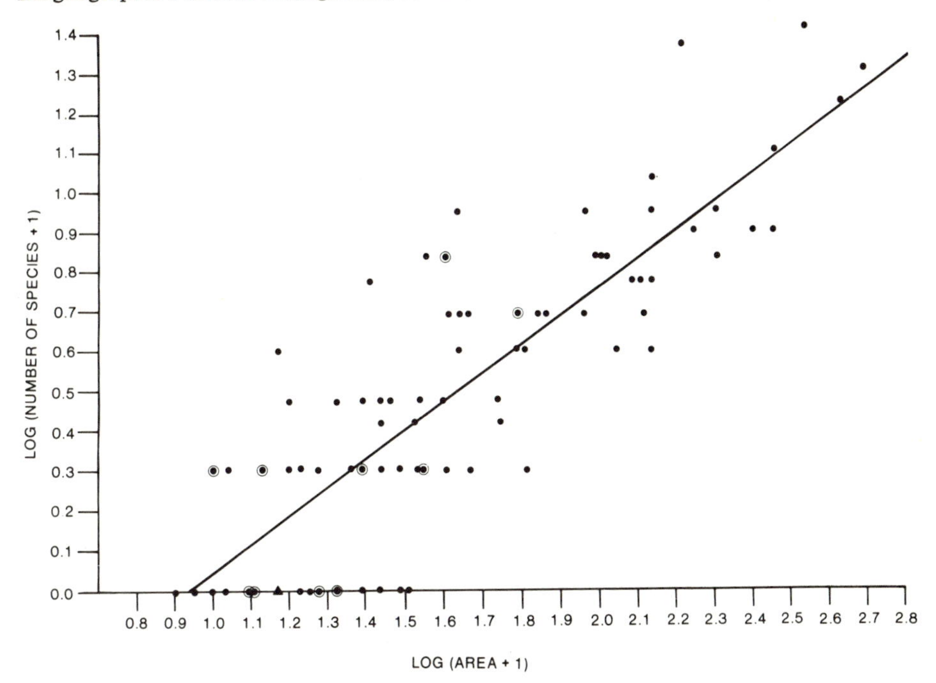

Figure 12-1. Species-area regression for insects on milkweeds from Cata, Venezuela. The regression equation is log (species +1) = −0.6700 + 0.7174 • log (area +1). Two points are indicated by dots with surrounding circles and 3 points by the triangle.

but the Cata species richness is higher than that of either Guadeloupe (H = 10.4846, $0.005 > P > 0.001$) or Maryland (H = 9.6747, $0.005 > P > 0.001$).

There is no significant difference of species per leaf among the four populations (Table 12-6; $F_s = 0.5530$, $0.75 > P > 0.50$).

There is a significant difference in the number of individuals per species of insect among the four populations (Table 12-7; H = 37.7582, $P < 0.001$). This difference is due to the large number of coccids present on the Guadeloupe and Ocumare milkweeds. Some of those plants had hundreds or in one case thousands of individual coccids, thus drastically increasing the average number of individuals per insect species.

The diversity of insects as measured by the Brillouin index is significantly different among the four milkweed populations (Table 12-8, H = 11.7923, $P < 0.001$). There is a higher average species diversity in Cata when compared with Guadeloupe (H = 13.2510, $P < 0.001$) and higher species diversity in Cata when compared with Maryland (H = 20.9110, $P < 0.001$). Tropical mainland insect diversity is greater than either temperate mainland or tropical island diversity.

The regression of evenness (J) on area is not statistically significant for the Cata ($F_s = 0.9119$, $P > 0.75$) or Maryland ($F_s = 0.0147$, $P > 0.75$) milkweeds, but it is significant for both the Ocumare ($F_s = 12.5931$, $0.005 > P > 0.001$) and Guadeloupe ($F_s = 4.3930$, $0.05 > P > 0.025$) milkweeds.

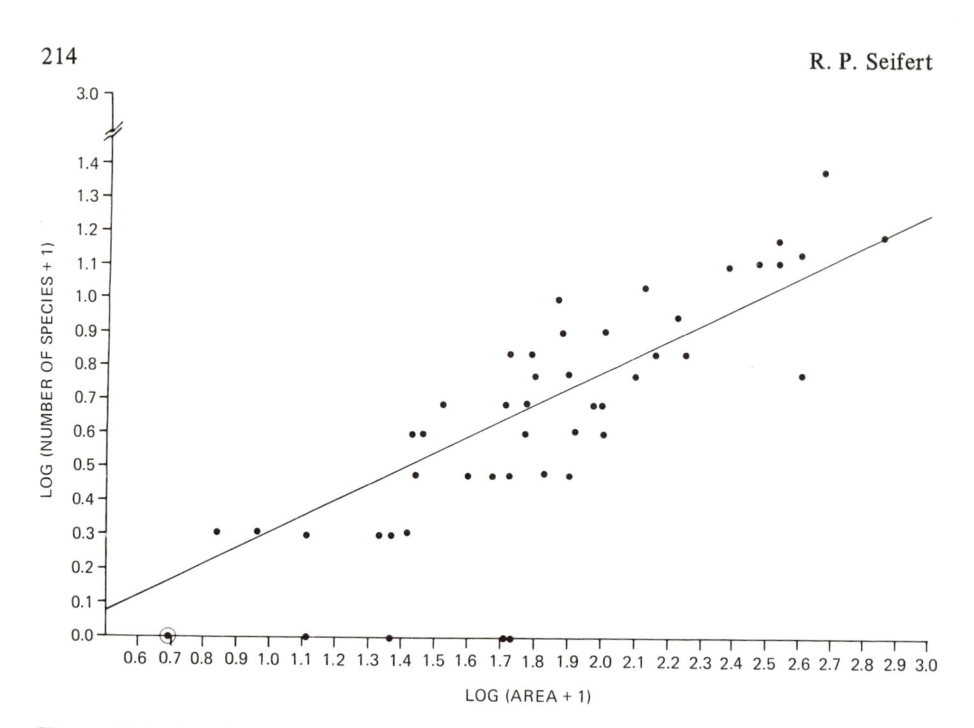

Figure 12-2. Species-area regression for insects on milkweeds from Ocumare, Venezuela. The regression equation is log (species +1) = –0.1687 + 0.4735 • log (area +1). Two points are indicated by the dot with the surrounding circle.

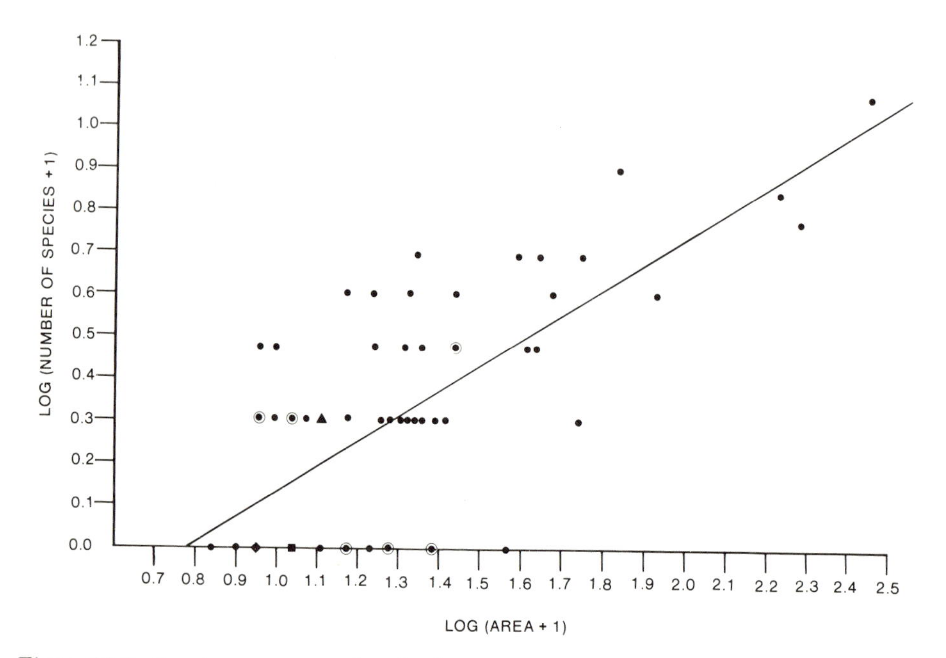

Figure 12-3. Species-area regression for insects on milkweeds from Bouillante, Guadeloupe. The regression equation is log (species +1) = –0.4666 + 0.5980 • log (area +1). Two points are indicated by dots with surrounding circles, 3 points by the triangle, 4 points by the square and 6 points by the diamond.

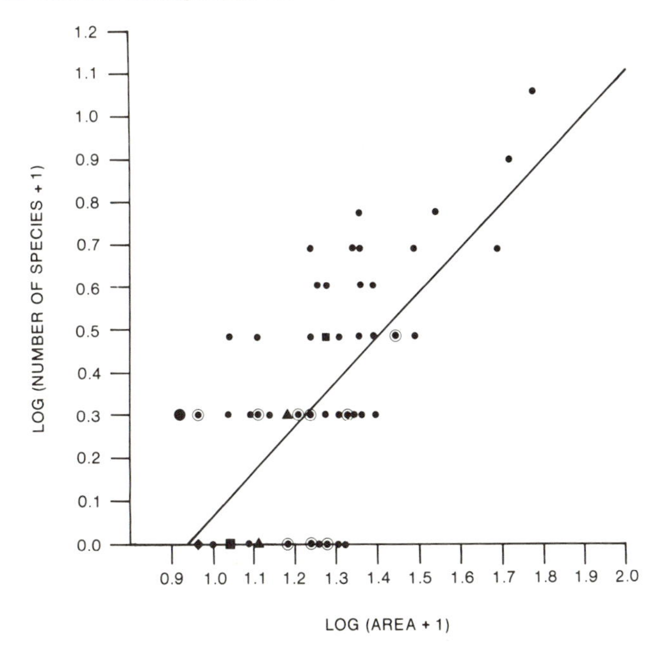

Figure 12-4. Species-area regression for insects on milkweeds from Capital Centre, Maryland. The regression equation is log (species +1) = −0.9827 + 1.0490 · log (area +1). Two points are indicated by dots with surrounding circles, 3 points by triangles, 4 points by squares and 7 points by the diamond.

4 Discussion

Insects on milkweed plants exhibit only some of the biogeographical patterns which have been shown for other organisms. All four milkweed insect communities exhibited increased number of insect species with increased area. However, the faunal coefficients associated with milkweeds as ecological islands for insect species are unusually high, ranging from 0.4735 for the Ocumare milkweed population to 1.049 for the Maryland milkweed population. These values lie above the range of the faunal coefficients (between 0.20 and 0.35) expected by MacArthur and Wilson (1967) for true islands. Strong (1974a), however, reported a faunal coefficient of 1.10 for insect species richness regressed against distributional area of British tree species. Other results of insect species richness regressed against (a) distributional area of California oak trees (Opler

Table 12-5. Mean, variance, median and range of the number of insect species per milkweed plant from four locations

	Location			
	Cata	Ocumare	Guadeloupe	Maryland
Mean	3.444	4.9600	1.5224	1.5000
Variance	19.4631	22.9371	3.8291	3.4314
Median	2.0	4.0	1.0	1.0
Range	22.0	23.0	11.0	11.0

Table 12-6. Mean and variance of the square root of the number of insect species per leaf on milkweed plants from four locations

	Location			
	Cata	Ocumare	Guadeloupe	Maryland
Mean	0.1910	0.2105	0.1834	0.2111
Variance	0.0156	0.0105	0.0242	0.0267

1974), (b) Atlantic region and California oak trees (Cornell and Washburn 1977), (c) area planted in cacao (Strong 1974b), (d) area planted in sugar cane (Strong et al. 1977) and (e) clump size of *Heliconia* inflorescences (Seifert 1975) all indicate faunal coefficients above those expected by MacArthur and Wilson (1967). Apparently, insect host plants typically have higher faunal coefficients than do true islands. Note that MacArthur and Wilson (1967) did not present a theoretical justification for a particular range of values of z, but simply reported an observational result. Indeed, various mechanisms to explain this result other than MacArthur and Wilson's (1967) immigration-extinction processes have been proposed (Conner and McCoy 1979, Strong 1979) and distributions yielding this kind of curve have been suggested to occur for nonbiological data (Anderson 1974) as well as island biogeographic data. The generalization of high z values for insects on host plants is contradicted by Lawton and Price (1979), who found a value of 0.139 for agromyzid fly larvae on British umbellifers, and by Raupp and Denno (1979), who found that sap-feeding insects living on patches of salt marsh grass showed no increase of species richness with grass patch size increase. In that study each species of a highly specialized guild of Homoptera and Hemiptera colonized each grass patch regardless of grass patch size. However, more recent studies (Denno et al., Chapter 9) indicate that, when smaller patch sizes are included, species richness increases with increasing area. High z values may also be associated with small true islands. Rusterholtz and Howe (1979) found that, on small true islands ranging in size from 0.04 to 17.2 ha in Burntside Lake in Minnesota, bird species richness had a z value of 0.44.

Evenness was correlated with area for the Guadeloupe and Ocumare milkweed populations. The decrease of evenness with area shown for these milkweed insects is similar to a decrease of evenness correlated with area which has been shown to occur for insects living in clumps of *Heliconia* inflorescences (Seifert 1975). In both cases the decrease in evenness was due to increasing population densities of the most common insects as the area of the plants increased. However, no evenness-area association

Table 12-7. Mean, variance, median and range of the number of individuals per species of insect on milkweed plants from four locations

	Location			
	Cata	Ocumare	Guadeloupe	Maryland
Mean	1.7475	16.3374	12.5065	0.9356
Variance	7.3085	2565.0466	3397.9071	1.4214
Median	1.0	2.8335	1.0	1.0
Range	20.5	327.9565	443.1	8.0

Table 12-8. Mean, variance, median and range of species diversity, measured by the Brillouin index of insects on milkweed plants from four locations

	Location			
	Cata	Ocumare	Guadeloupe	Maryland
Mean	0.2169	0.2358	0.0828	0.0964
Variance	0.0620	0.0435	0.0160	0.0236
Median	0.0	0.1938	0.0	0.0
Range	0.9295	0.6512	0.4310	0.7512

was shown to be significant for either of the other two milkweed populations, indicating that a decrease in insect evenness with increasing plant size is not a universal characteristic of host plants as ecological islands for insects.

Nearest neighbor distance effects were not shown to be significant in determining the number of species of insects on individual milkweed plants. Milkweed plants do not follow an island pattern of increased species with increased proximity to other islands or to a source area. Distance effects have also been shown to be unimportant in determining species richness of insects in clumps of *Heliconia* inflorescences (Seifert 1975). Distance may be generally unimportant for insects on host plants over small geographic distributions, since there is not usually a source area from which insect species can be drawn. While certain kinds of milkweed plants, particularly those with inflorescences or pods, may serve as reservoirs for some insects such as *Oncopeltus* bug species (Root and Chaplin 1976, Ralph 1977), the milkweeds in a given deme are close enough so that dispersal among plants is not limited by distance.

The diversity of insects on milkweed plants shows that a major biogeographic pattern is superimposed on the ecological island patterns presented by insect host plants. Species diversity, as estimated by the Brillouin index, is higher in tropical mainland areas than in the tropical island location. However, the tropical island and temperate mainland diversities seem comparable. Other studies on the diversity of Neotropical and Nearctic insects contrasting mainland and island patterns have concluded that tropical mainland insect communities are the most diverse of the three kinds of communities (Janzen 1973b), and that insect diversities on tropical islands are similar to insect diversities in temperate mainland areas (Allan et al. 1973, Culver 1974). This milkweed study is unique, since it considers a group of insects the majority of which have coevolved with plants of apparently unusual toxicity (Brower and Brower 1964, Brower 1969). Further, large population densities of a few organisms are common attributes of islands due either to competitive release or to low levels of predation (MacArthur and Wilson 1967, Janzen 1968, 1973a, Carlquist 1974, Simberloff 1974). This pattern is shown by the Guadeloupe milkweed insects, where the number of individuals per insect species is high due to the density of coccids on some of the milkweed individuals.

In conclusion, milkweeds do act as ecological islands for their insect hosts, but ecological island patterns in general are different from true island patterns. Host plants as ecological islands have faunal coefficients which are higher than those of true islands and do not show distance effects. Insect host plants follow the general biogeographic pattern of showing highest species diversity and species richness in tropical mainland locations. Slobodkin and Sanders (1969) and Futuyma (1973, 1976) have postulated that high tropical diversity is generated from a large number of coevolved organisms showing greater degrees of specialization than the organisms in temperate regions. Based on my study, insect species diversity in the tropical mainland areas is due at least partially to the additional number of species which can exist on a single food plant. More insect species exist on each tropical mainland milkweed plant than do on each temperate mainland or tropical island milkweed plant, even though the number of insect species per leaf is similar in all four locations.

While this paper documents various island biogeography and species diversity phenomena, it also brings into question the importance of these kinds of studies. Species-area curves can be generated by variety of mechanisms, not all of which have biological meaning (Anderson 1974, Conner and McCoy 1979, Lawton and Price 1979, Strong 1979). Conner and McCoy (1979) suggest that at least three different kinds of area effects may produce the standard species-area curves. Lawton and Price (1979) have shown how area accounts for only 32% of the variance about the regression line of agromyzid fly larvae on British umbellifers. Gould (1979) points out that species-area curves have the same form as power-function curves of allometric growth. Further, in a paper which seems to have been largely overlooked by proponents of species-area curves, Anderson (1974) has shown that such distributions as frequencies of atoms of elements in the solar system, species per genus (for many organisms) and even frequencies of letters on a printed page all fall into the same family of curves as species-area curves and can be approximated by his equation:

$$\log y = \log c - m \cdot \log x \qquad \text{(Anderson 1974, p. 329).}$$

The documentation of a range of faunal coefficients (z values) for different taxonomic groups and different geographic regions or the difference between true islands and ecological islands may generate interesting contrasts, as they have done in this paper. However, critical tests of the mechanisms which generate these curves are only infrequently made (Seifert 1975, Simberloff 1976, Lawton and Price 1979). For example, elsewhere (Seifert 1975) I was able to show that differential sequences in inflorescence blooming between two species of *Heliconia* plants led to different ways in which these plants accumulated insect species. The *Heliconia* species with staggered blooming allowed for a wider variety of different inflorescence specialists in each *Heliconia* clump, which resulted in a higher z value than that associated with the *Heliconia* species, which showed temporally uniform blooming. I suspect that more mechanistic studies will prove of increasingly greater value, now that different ranges of faunal coefficients for different kinds of biological processes and geographic divisions have been shown to exist.

Acknowledgments. I am pleased to thank the many people who helped in the production of this manuscript. I was aided in the field in Venezuela by Guido Pereira S. and by Mercedes Asclapés in 1977 and by Nancy J. Beaman in 1979. Carlos Machado Allison and the Instituto de Zoología Tropical, Universidad Central de Venezuela (Caracas) provided field transportation in 1979. F. Javier Castillo provided lodging in Caracas, Venezuela, and Francisco Fernández Yépez allowed me to use the facilities of

Estación Biológica Rancho Grande while in Venezuela. I thank Mr. Abe Pollin and the Washington Bullets for allowing me to conduct research at Capital Centre. This manuscript has profited from discussions with F. Javier Castillo, Hugh Dingle, J. Steven Farris, Douglas Gill, Charles Mitter, Kittie F. Parker, Peter W. Price and Florence Hammett Seifert. Field expenses were covered by The George Washington University Committee on Research (1977, 1978) and National Science Foundation Grant DEB 79-06593 (1979).

This paper is dedicated to Dr. Kenneth A. Christiansen of Grinnell College who, while I was an undergraduate, helped to stimulate and mold my interests in both entomology and community ecology.

References

Allan, J. D., Barnthose, L. W., Prestby, R. A., Strong, D. R., Jr.: On foliage arthropod communities of Puerto Rican second growth vegetation. Ecology 54, 628-632 (1973).

Anderson, S.: Patterns of faunal evolution. Q. Rev. Biol. 49, 311-332 (1974).

Brower, L. P.: Ecological chemistry. Sci. Am. 222, 22-29 (1969).

Brower, L. P., Brower, J. V. Z.: Birds, butterflies, and plant poisons: a study in ecological chemistry. Zoologica 49, 137-159 (1964).

Carlquist, S.: Island Biology. New York: Columbia Univ. Press, 1974.

Conner, E. F., McCoy, E. D.: The statistics and biology of species-area relationship. Am. Nat. 113, 791-833 (1979).

Cornell, H. V., Washburn, J. O.: Evolution of the richness-area correlation for cynipid gall wasps on oak trees: a comparison of two geographic areas. Evolution 33, 257-274 (1979).

Culver, D. C.: Species packing in Caribbean and north temperate ant communities. Ecology 55, 974-988 (1974).

Darwin, C.: The Origin of Species by Means of Natural Selection. London: Murray, 1859.

Futuyma, D. J.: Community structure and stability in constant environments. Am. Nat. 107, 443-445 (1973).

Futuyma, D. J.: Food plant specialization and environmental predictability in Lepidoptera. Am. Nat. 110, 285-292 (1976).

Gould, S. J.: An allometric interpretation of species-area curves: the meaning of the coefficient. Am. Nat. 114, 335-343 (1979).

Holdridge, L. R.: Life-zone Ecology. San Jose, Costa Rica: Tropical Science Center, 1964.

Imms, A. D.: A General Textbook of Entomology. London: Methuen, 1957.

Janzen, D. H.: Host plants as islands in evolutionary and contemporary time. Am. Nat. 102, 592-595 (1968).

Janzen, D. H.: Host plants as islands. II. Competition in evolutionary and contemporary time. Am. Nat. 107, 786-790 (1973a).

Janzen, D. H.: Sweep samples of tropical foliage insects: effects of seasons, vegetation types, elevation, time of day, and insularity. Ecology 54, 687-708 (1973b).

Lawton, J. H., Price, P. W.: Species richness of parasites on hosts: agromyzid flies on the British Umbelliferae. J. Anim. Ecol. 48, 619-637 (1979).

MacArthur, R. H.: Patterns of species diversity. Biol. Rev. 40, 510-533 (1965).

MacArthur, R. H.: Patterns of communities in the tropics. Biol. J. Linn. Soc. 1, 19-30 (1969).

MacArthur, R. H.: Geographical Ecology: Patterns in the Distribution of Species. New York: Harper and Row, 1972.

MacArthur, R. H., Wilson, E. O.: An equilibrium theory of insular zoogeography. Evolution 17, 373-387 (1963).

MacArthur, R. H., Wilson, E. O.: The Theory of Island Biogeography. Princeton, N.J.: Princeton Univ. Press, 1967.

Opler, P. A.: Oaks as evolutionary islands for leaf-mining insects. Am. Sci. 62, 67-73 (1974).

Pianka, E. R.: Latitudinal gradients in species diversity: a review of concepts. Am. Nat. 100, 33-46 (1966).

Pianka, E. R.: Evolutionary Ecology. New York: Harper and Row, 1978.

Pielou, E. C.: Population and Community Ecology. New York: Gordon and Breach, 1974.

Price, P. W.: Insect Ecology. New York: Wiley, 1975.

Ralph, C. P.: Effect of host plant density on populations of a specialized, seed-sucking bug, *Oncopeltus fasciatus*. Ecology 58, 799-809 (1977).

Raupp, M. J., Denno, R. F.: The influence of patch size on a guild of sap-feeding insects that inhabit the salt marsh grass *Spartina patens*. Environ. Entomol. 8, 412-417 (1979).

Richards, P. W.: The Tropical Rain Forest. New York: Cambridge Univ. Press, 1952.

Root, R. B., Chaplin, S. J.: The life-styles of tropical milkweed bugs, *Oncopeltus* (Hemiptera: Lygaeidae) utilizing the same hosts. Ecology 57, 132-140 (1976).

Rusterholtz, K. A., Howe, R. W.: Species-area relations of birds on small islands in a Minnesota lake. Evolution 33, 468-477 (1979).

Seifert, R. P.: Clumps of *Heliconia* inflorescences as ecological islands. Ecology 56, 1416-1422 (1975).

Simberloff, D. S.: Equilibrium theory of island biogeography and ecology. Ann. Rev. Ecol. Syst. 5, 161-182 (1974).

Simberloff, D. S.: Experimental zoogeography of islands: effects of island size. Ecology 57, 629-648 (1976).

Slobodkin, L. B., Sanders, H. L.: On the contribution of environmental predictability to species diversity. Brookhaven Symp. Biol. 22, 82-95 (1969).

Strong, D. R., Jr.: Nonasymptotic species richness models and the insects of British trees. Proc. Nat. Acad. Sci. USA 71, 2766-2769 (1974a).

Strong, D. R., Jr.: Rapid asymptotic species accumulation in phytophagous insect communities: the pests of cacao. Science 185, 1064-1066 (1974b).

Strong, D. R., Jr.: Biogeographic dynamics of insect-host plant communities. Ann. Rev. Entomol. 24, 89-119 (1979).

Strong, D. R., Jr., McCoy, E. D., Rey, J. R.: Time and the number of herbivore species: the pests of sugarcane. Ecology 58, 167-175 (1977).

Wallace, A. E.: The Geographic Distribution of Animals (2 Vols.). New York: Hafner, 1876.

Index